# Suzuki GP100 & 125 Owners Workshop Manual

## by Chris Rogers
with an additional Chapter on the 1985 to 1986 GP125 N model
## by Pete Shoemark

**Models covered**
GP100 UN. 98cc. February 1980 to September 1982
GP100 UX. 98cc. September 1982 to June 1983
GP100 UD. 98cc. June 1983 to April 1991
GP100 UL. 98cc. April 1991 to April 1993
GP100 C. 98cc. August 1978 to August 1980
GP100 N. 98cc. August 1980 to March 1981
GP100 X. 98cc. March 1981 to March 1984
GP100 ED. 98cc. March 1983 to March 1986
GP125 C. 123cc. February 1978 to October 1979
GP125 N. 123cc. October 1979 to July 1982, 1985 to 1986
GP125 X. 123cc. July 1982 to September 1983
GP125 D. 123cc. September 1983 to February 1989

**ISBN 978 1 85010 929 7**

Printed in the UK          (576 - 5S7)

ABCDE
F

2

**Haynes Publishing Group**
Sparkford Nr Yeovil
Somerset BA22 7JJ England

**Haynes Publications, Inc**
859 Lawrence Drive
Newbury Park
California 91320 USA

**British Library Cataloguing in Publication Data**

A catalogue record for this book is available from the British Library

# Acknowledgements

Our thanks are due to Fran Ridewood and Co, of Wells, Somerset who supplied the Suzuki GP125 model featured in the photographs throughout the manual, and who also supplied technical information relating to the Suzuki GP100 and GP125 models.

We should also like to thank Heron Suzuki (GB) Ltd for permission to use their line illustrations, and members of the Technical Service Department of that company who gave advice and checked the contents of the manual for technical accuracy. The GP100 model featured on the front cover of this manual was supplied by Paul Branson Motorcycles of Yeovil, Somerset.

Finally, we would like to thank the Avon Rubber Company, who kindly supplied information and technical assistance on tyre fitting; NGK Spark Plugs (UK) Ltd. for information on spark plug maintenance and electrode conditions, and Renold Ltd. for advice on chain care and renewal.

# About this manual

The purpose of this manual is to present the owner with a concise and graphic guide which will enable him to tackle any operation from basic routine maintenance to a major overhaul. It has been assumed that any work would be undertaken without the luxury of a well-equipped workshop and a range of manufacturer's service tools.

To this end, the machine featured in the manual was stripped and rebuilt in our own workshop, by a team comprising a mechanic, a photographer and the author. The resulting photographic sequence depicts events as they took place, the hands shown being those of the author and the mechanic.

The use of specialised, and expensive, service tools was avoided unless their use was considered to be essential due to risk of breakage or injury. There is usually some way of improvising a method of removing a stubborn component, provided that a suitable degree of care is exercised.

The author learnt his motorcycle mechanics over a number of years, faced with the same difficulties and using similar facilities to those encountered by most owners. It is hoped that this practical experience can be passed on through the pages of this manual.

Where possible, a well-used example of the machine is chosen for the workshop project, as this highlights any areas which might be particularly prone to giving rise to problems. In this way, any such difficulties are encountered and resolved before the text is written, and the techniques used to deal with them can be incorporated in the relevant sections. Armed with a working knowledge of the machine, the author undertakes a considerable amount of research in order that the maximum amount of data can be included in this manual.

Each Chapter is divided into numbered sections. Within these sections are numbered paragraphs. Cross reference throughout the manual is quite straightforward and logical. When reference is made 'See Section 6.10' it means Section 6, paragraph 10 in the same Chapter. If another Chapter were intended the reference would read, for example, 'See Chapter 2, Section 6.10'. All the photographs are captioned with a section/paragraph number to which they refer and are relevant to the Chapter text adjacent.

Figures (usually line illustrations) appear in a logical but numerical order, within a given Chapter. Fig. 1.1 therefore refers to the first figure in Chapter 1.

Left-hand and right-hand dscriptions of the machines and their components refer to the left and right of a given machine when the rider is seated normally.

Motorcycle manufacturers continually make changes to specifications and recommendations, and these, when notified, are incorporated into our manuals at the earliest opportunity.

# Contents

Right-hand view of the Suzuki GP100 U

Right-hand view of the Suzuki GP125

Close-up of the GP100 U engine and gearbox unit

# Introduction to the Suzuki GP100 and GP125

Allthough the Suzuki Motor Company Limited commenced manufacturing motorcycles as early as 1936, it was not until 1963 that their machines were first imported into the UK. The first of the twin cylinder models to be imported, the T10, became available during 1964 and it was immediately obvious that this particular model would be well received by holders of a provisional driving licence, who were at that time restricted to an engine capacity limit of 250 cc. This was due to the fact that not many 250 cc motorcycles of the time were capable of a genuine near 90 mph maximum speed and yet able to show a fuel consumption figure well in excess of 100 mpg.

The T10 was closely followed by the T20 Super Six which proved to be even more of a success as a sports machine and was so named because of its six-speed gearbox. In an attempt to gain sales in the commuter section of the market, Suzuki also released the B100P which incorporated a conventional four-port, single cylinder, two-stroke engine. This machine, affectionately renamed the 'Bloop' by those who rode it, gained a good reputation and sold well, thus providing encouragement for Suzuki to refine the design of the engine and frame components. This they did, and in 1969 the A100 model was produced. This machine, as well as including the 'Posi-Force' (CCI) system of lubrication developed by Suzuki and incorporated in the B 100 P model, featured the use of a crankshaft-mounted disc to control induction. Further progress has led to an engine of similar design being incorporated in a lightweight frame of tubular steel construction (as opposed to the earlier spine, pressed steel construction) and it is this machine, the GP100, which is featured in this Manual, along with its larger counterpart, the GP125.

The design differences between GP100 and GP125 models of the same year are minimal, the one obvious difference being the change in cylinder bore diameter to account for the difference in engine cubic capacity between the model types. The first model to be introduced into the UK market was the GP125 C of February, 1978; this being followed by the GP100 C in August of the same year. In February of 1980, a 'basic' version of the GP100 was introduced. Known as the GP100 U, this machine was presented as a basic commuter machine and differed from the standard GP100 model in having a drum brake fitted to the front wheel instead of the sportier disc brake. Other differences were minor and mainly of the 'cost saving' cosmetic type, such as the painting instead of chroming of various cycle components, including the mudguards.

October 1979 saw the introduction of the GP125 N model and August 1980, the introduction of the GP100 N model, the only differences between these models and those of the previous year being cosmetic. The GP125 N was superseded by the GP125 X in 1982, which remained in production until 1983 when the GP125 D was introduced. The GP100 N changed to the GP100 X in 1981 and then became the GP100 D in 1983. In order to give this machine a more sporty image than the still produced GP100 U model, design changes were made which included the fitting of cast alloy wheels of Suzuki's own design.

# Model dimensions and weight

|  | GP100 | GP125 |
| --- | --- | --- |
| Overall length | 1900 mm (74.8 in) | 1905 mm (75.0 in) |
| Overall width | 750 mm (29.5 in) | 750 mm (29.5 in) |
| Overall height | 1045 mm (41.1 in) | 1075 mm (42.3 in) |
| Wheelbase | 1225 mm (48.2 in) | 1230 mm (48.4 in) |
| Ground clearance | 145 mm (5.7 in) | 150 mm (5.9 in) |
| Seat height | 770 mm (30.3 in) | 770 mm (30.3 in) |
| Dry weight: | | |
|     U suffix models | 86 kg (190 lb) | 89 kg (196 lb) |
|     All other models | 89 kg (196 lb) | 92 kg (203 lb) |

# Ordering spare parts

When ordering spare parts for any Suzuki, it is advisable to deal direct with an official Suzuki agent who should be able to supply most of the parts ex stock. Parts cannot be obtained from Suzuki direct even if the parts required are not held in stock. Always, quote the engine and frame numbers in full, especially if parts are required for earlier models. The engine number is stamped on the top surface of the left-hand crankcase half. The frame number is stamped on the right-hand side of the steering head.

Use only genuine Suzuki spares. Some pattern parts are available that are made in Japan and may be packed in similar looking packages. They should only be used if genuine parts are hard to obtain or in an emergency, for they do not normally last as long as genuine parts, even although there may be a price advantage.

Some of the more expendable parts such as spark plugs, bulbs, oils and greases etc., can be obtained from accessory shops and motor factors, who have convenient opening hours, and can often be found not far from home. It is also possible to obtain parts on a Mail Order basis from a number of specialists who advertise regularly in the motorcycle magazines.

Location of frame number

Location of engine number

# Safety first!

Professional motor mechanics are trained in safe working procedures. However enthusiastic you may be about getting on with the job in hand, do take the time to ensure that your safety is not put at risk. A moment's lack of attention can result in an accident, as can failure to observe certain elementary precautions.

There will always be new ways of having accidents, and the following points do not pretend to be a comprehensive list of all dangers; they are intended rather to make you aware of the risks and to encourage a safety-conscious approach to all work you carry out on your vehicle.

### Essential DOs and DON'Ts

**DON'T** start the engine without first ascertaining that the transmission is in neutral.

**DON'T** suddenly remove the filler cap from a hot cooling system – cover it with a cloth and release the pressure gradually first, or you may get scalded by escaping coolant.

**DON'T** attempt to drain oil until you are sure it has cooled sufficiently to avoid scalding you.

**DON'T** grasp any part of the engine, exhaust or silencer without first ascertaining that it is sufficiently cool to avoid burning you.

**DON'T** allow brake fluid or antifreeze to contact the machine's paintwork or plastic components.

**DON'T** syphon toxic liquids such as fuel, brake fluid or antifreeze by mouth, or allow them to remain on your skin.

**DON'T** inhale dust – it may be injurious to health (see *Asbestos* heading).

**DON'T** allow any spilt oil or grease to remain on the floor – wipe it up straight away, before someone slips on it.

**DON'T** use ill-fitting spanners or other tools which may slip and cause injury.

**DON'T** attempt to lift a heavy component which may be beyond your capability – get assistance.

**DON'T** rush to finish a job, or take unverified short cuts.

**DON'T** allow children or animals in or around an unattended vehicle.

**DON'T** inflate a tyre to a pressure above the recommended maximum. Apart from overstressing the carcase and wheel rim, in extreme cases the tyre may blow off forcibly.

**DO** ensure that the machine is supported securely at all times. This is especially important when the machine is blocked up to aid wheel or fork removal.

**DO** take care when attempting to slacken a stubborn nut or bolt. It is generally better to pull on a spanner, rather than push, so that if slippage occurs you fall away from the machine rather than on to it.

**DO** wear eye protection when using power tools such as drill, sander, bench grinder etc.

**DO** use a barrier cream on your hands prior to undertaking dirty jobs – it will protect your skin from infection as well as making the dirt easier to remove afterwards; but make sure your hands aren't left slippery. Note that long-term contact with used engine oil can be a health hazard.

**DO** keep loose clothing (cuffs, tie etc) and long hair well out of the way of moving mechanical parts.

**DO** remove rings, wristwatch etc, before working on the vehicle – especially the electrical system.

**DO** keep your work area tidy – it is only too easy to fall over articles left lying around.

**DO** exercise caution when compressing springs for removal or installation. Ensure that the tension is applied and released in a controlled manner, using suitable tools which preclude the possibility of the spring escaping violently.

**DO** ensure that any lifting tackle used has a safe working load rating adequate for the job.

**DO** get someone to check periodically that all is well, when working alone on the vehicle.

**DO** carry out work in a logical sequence and check that everything is correctly assembled and tightened afterwards.

**DO** remember that your vehicle's safety affects that of yourself and others. If in doubt on any point, get specialist advice.

**IF,** in spite of following these precautions, you are unfortunate enough to injure yourself, seek medical attention as soon as possible.

### Asbestos

Certain friction, insulating, sealing, and other products – such as brake linings, clutch linings, gaskets, etc – contain asbestos. *Extreme care must be taken to avoid inhalation of dust from such products since it is hazardous to health.* If in doubt, assume that they *do* contain asbestos.

### Fire

Remember at all times that petrol (gasoline) is highly flammable. Never smoke, or have any kind of naked flame around, when working on the vehicle. But the risk does not end there – a spark caused by an electrical short-circuit, by two metal surfaces contacting each other, by careless use of tools, or even by static electricity built up in your body under certain conditions, can ignite petrol vapour, which in a confined space is highly explosive.

Always disconnect the battery earth (ground) terminal before working on any part of the fuel or electrical system, and never risk spilling fuel on to a hot engine or exhaust.

It is recommended that a fire extinguisher of a type suitable for fuel and electrical fires is kept handy in the garage or workplace at all times. Never try to extinguish a fuel or electrical fire with water.

**Note:** *Any reference to a 'torch' appearing in this manual should always be taken to mean a hand-held battery-operated electric lamp or flashlight. It does **not** mean a welding/gas torch or blowlamp.*

### Fumes

Certain fumes are highly toxic and can quickly cause unconsciousness and even death if inhaled to any extent. Petrol (gasoline) vapour comes into this category, as do the vapours from certain solvents such as trichloroethylene. Any draining or pouring of such volatile fluids should be done in a well ventilated area.

When using cleaning fluids and solvents, read the instructions carefully. Never use materials from unmarked containers – they may give off poisonous vapours.

Never run the engine of a motor vehicle in an enclosed space such as a garage. Exhaust fumes contain carbon monoxide which is extremely poisonous; if you need to run the engine, always do so in the open air or at least have the rear of the vehicle outside the workplace.

### The battery

Never cause a spark, or allow a naked light, near the vehicle's battery. It will normally be giving off a certain amount of hydrogen gas, which is highly explosive.

Always disconnect the battery earth (ground) terminal before working on the fuel or electrical systems.

If possible, loosen the filler plugs or cover when charging the battery from an external source. Do not charge at an excessive rate or the battery may burst.

Take care when topping up and when carrying the battery. The acid electrolyte, even when diluted, is very corrosive and should not be allowed to contact the eyes or skin.

If you ever need to prepare electrolyte yourself, always add the acid slowly to the water, and never the other way round. Protect against splashes by wearing rubber gloves and goggles.

### Mains electricity and electrical equipment

When using an electric power tool, inspection light etc, always ensure that the appliance is correctly connected to its plug and that, where necessary, it is properly earthed (grounded). Do not use such appliances in damp conditions and, again, beware of creating a spark or applying excessive heat in the vicinity of fuel or fuel vapour. Also ensure that the appliances meet the relevant national safety standards.

### Ignition HT voltage

A severe electric shock can result from touching certain parts of the ignition system, such as the HT leads, when the engine is running or being cranked, particularly if components are damp or the insulation is defective. Where an electronic ignition system is fitted, the HT voltage is much higher and could prove fatal.

# Working conditions and tools

When a major overhaul is contemplated, it is important that a clean, well-lit working space is available, equipped with a workbench and vice, and with space for laying out or storing the dismantled assemblies in an orderly manner where they are unlikely to be disturbed. The use of a good workshop will give the satisfaction of work done in comfort and without haste, where there is little chance of the machine being dismantled and reassembled in anything other than clean surroundings. Unfortunately, these ideal working conditions are not always practicable and under these latter circumstances when improvisation is called for, extra care and time will be needed.

The other essential requirement is a comprehensive set of good quality tools. Quality is of prime importance since cheap tools will prove expensive in the long run if they slip or break when in use, causing personal injury or expensive damage to the component being worked on. A good quality tool will last a long time, and more than justify the cost.

For practically all tools, a tool factor is the best source since he will have a very comprehensive range compared with the average garage or accessory shop. Having said that, accessory shops often offer excellent quality tools at discount prices, so it pays to shop around. There are plenty of tools around at reasonable prices, but always aim to purchase items which meet the relevant national safety standards. If in doubt, seek the advice of the shop proprietor or manager before making a purchase.

The basis of any tool kit is a set of open-ended spanners, which can be used on almost any part of the machine to which there is reasonable access. A set of ring spanners makes a useful addition, since they can be used on nuts that are very tight or where access is restricted. Where the cost has to be kept within reasonable bounds, a compromise can be effected with a set of combination spanners – open-ended at one end and having a ring of the same size on the other end. Socket spanners may also be considered a good investment, a basic $3/8$ in or $1/2$ in drive kit comprising a ratchet handle and a small number of socket heads, if money is limited. Additional sockets can be purchased, as and when they are required. Provided they are slim in profile, sockets will reach nuts or bolts that are deeply recessed. When purchasing spanners of any kind, make sure the correct size standard is purchased. Almost all machines manufactured outside the UK and the USA have metric nuts and bolts, whilst those produced in Britain have BSF or BSW sizes. The standard used in USA is AF, which is also found on some of the later British machines. Others tools that should be included in the kit are a range of crosshead screwdrivers, a pair of pliers and a hammer.

When considering the purchase of tools, it should be remembered that by carrying out the work oneself, a large proportion of the normal repair cost, made up by labour charges, will be saved. The economy made on even a minor overhaul will go a long way towards the improvement of a toolkit.

In addition to the basic tool kit, certain additional tools can prove invaluable when they are close to hand, to help speed up a multitude of repetitive jobs. For example, an impact screwdriver will ease the removal of screws that have been tightened by a similar tool, during assembly, without a risk of damaging the screw heads. And, of course, it can be used again to retighten the screws, to ensure an oil or airtight seal results. Circlip pliers have their uses too, since gear pinions, shafts and similar components are frequently retained by circlips that are not too easily displaced by a screwdriver. There are two types of circlip pliers, one for internal and one for external circlips. They may also have straight or right-angled jaws.

One of the most useful of all tools is the torque wrench, a form of spanner that can be adjusted to slip when a measured amount of force is applied to any bolt or nut. Torque wrench settings are given in almost every modern workshop or service manual, where the extent to which a complex component, such as a cylinder head, can be tightened without fear of distortion or leakage. The tightening of bearing caps is yet another example. Overtightening will stretch or even break bolts, necessitating extra work to extract the broken portions.

As may be expected, the more sophisticated the machine, the greater is the number of tools likely to be required if it is to be kept in first class condition by the home mechanic. Unfortunately there are certain jobs which cannot be accomplished successfully without the correct equipment and although there is invariably a specialist who will undertake the work for a fee, the home mechanic will have to dig more deeply in his pocket for the purchase of similar equipment if he does not wish to employ the services of others. Here a word of caution is necessary, since some of these jobs are best left to the expert. Although an electrical multimeter of the AVO type will prove helpful in tracing electrical faults, in inexperienced hands it may irrevocably damage some of the electrical components if a test current is passed through them in the wrong direction. This can apply to the synchronisation of twin or multiple carburettors too, where a certain amount of expertise is needed when setting them up with vacuum gauges. These are, however, exceptions. Some instruments, such as a strobe lamp, are virtually essential when checking the timing of a machine powered by CDI ignition system. In short, do not purchase any of these special items unless you have the experience to use them correctly.

Although this manual shows how components can be removed and replaced without the use of special service tools (unless absolutely essential), it is worthwhile giving consideration to the purchase of the more commonly used tools if the machine is regarded as a long term purchase Whilst the alternative methods suggested will remove and replace parts without risk of damage, the use of the special tools recommended and sold by the manufacturer will invariably save time.

# SUZUKI GP100 & 125 SINGLES

## Check list

**Weekly or every 200 miles (300 km)**

1  Safety inspection
2  Check the operation of the lights and instruments
3  Check the tyre pressures (cold) and tyre condition
4  Check the level of engine oil in the oil tank and top up if necessary
5  Lubricate the exposed portions of the control cables
6  Lubricate the final drive chain
7  Check the level of hydraulic fluid in the front brake reservoir (disc brake models)

**Monthly or every 600 miles (1000 km)**

1  Check and adjust the final drive chain
2  Check the level of electrolyte in the battery

**Three monthly or every 2000 miles (3000 km)**

1  Remove and regrease the final drive chain
2  Lubricate the control cables
3  Lubricate all pivot points and check the tightness of all fasteners
4  Check and adjust the oil pump setting
5  Change the gearbox oil
6  Adjust the carburettor
7  Adjust the throttle cable
8  Clean the air filter element
9  Check the spark plug gap
10  Clean, check and adjust the contact breaker points
11  Check the ignition timing
12  Check and adjust the clutch
13  Ensure that the cylinder head nuts are tightened to the correct torque setting
14  Check the front brake pads for wear (disc brake models)
15  Check and adjust the front brake cable (drum brake models)
16  Check and adjust the rear brake
17  Check the degree of brake shoe wear

**Six monthly or every 4000 miles (6000 km)**

1  Renew the spark plug
2  Clean the fuel tap filter element
3  Decarbonise the cylinder head and barrel
4  Check and adjust the steering head bearings
5  Lubricate the speedometer and tachometer cables
6  Lubricate the brake operating cam shaft (drum brakes)
7  Check the wheels
8  Check the suspension

**Yearly or every 8000 miles (12 000 km)**

1  Dismantle and clean the carburettor
2  Renew the contact breaker points
3  Change the front fork oil
4  Grease the wheel bearings and speedometer drive
5  Renew the brake fluid

**Two yearly or every 16 000 miles (24 000 km)**

1  Renew the fuel feed pipe
2  Renew the hydraulic brake hose (disc brake models)
3  Grease the steering head and swinging arm pivot bearings

## Adjustment data

| Tyre pressures GP100 | Solo | With pillion |
|---|---|---|
| Front | 25 psi (1.75 kg/cm²) | 25 psi (1.75 kg/cm²) |
| Rear | 28 psi (2.00 kg/cm²) | 32 psi (2.25 kg/cm²) |
| GP125 | | |
| Front | 25 psi (1.75 kg/cm²) | 25 psi (1.75 kg/cm²) |
| Rear | 32 psi (2.25 kg/cm²) | 36 psi (2.50 kg/cm²) |

| | |
|---|---|
| Sparking plug gap | 0.6 - 0.8 mm (0.024 - 0.031 in) |
| Sparking plug type | NGK B8HS or ND W24FS |
| Contact breaker gap | 0.35 mm (0.014 in) |
| Ignition timing | 20° ± 2° BTDC |
| Idle speed | 1300 ± 150 rpm |

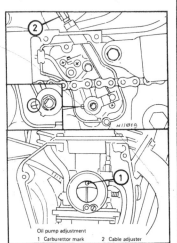

Oil pump adjustment
1  Carburettor mark    2  Cable adjuster

## Recommended lubricants

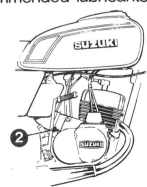

| Component | Quantity | Grade |
|---|---|---|
| 1 Engine oil | 1.2 lit (2.3 Imp pint) | Suziki CCI, CCI Super or equivalent 2-stroke oil |
| 2 Transmission oil: Dry | 850 cc (1.50 Imp pint) | SAE 20W/40 |
| At oil change | 800 cc (1.41 Imp pint) | |
| 3 Front forks (per leg) | 90 cc (3.17 Imp fl oz) | SAE 10W/20 |
| 4 Final drive chain | As required | Aerosol chain lubricant |
| 5 Wheel bearings | As required | High melting point grease |
| 6 Steering head bearings | As required | High melting point grease |
| 7 Pivot points | As required | High melting point grease |
| 8 Control cables | As required | Light machine oil |
| 9 Hydraulic disc brake | As required | SAE J1703 hydraulic fluid |

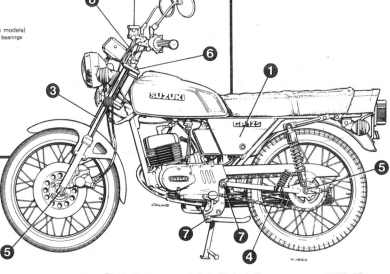

# ROUTINE MAINTENANCE GUIDE

# Routine maintenance

Periodic routine maintenance is essential to keep the motorcycle in a peak and safe condition. Routine maintenance also saves money because it provides the opportunity to detect and remedy a fault before it develops further and causes more damage. Maintenance should be undertaken on either a calendar or mileage basis depending on whichever comes sooner. The period between maintenance tasks serves only as a guide since there are many variables eg; age of machine, riding technique and adverse conditions.

The maintenance instructions are generally those recommended by the manufacturer but are supplemented by additional tasks which, through practical experience, the author recommends should be carried out at the intervals suggested. The additional tasks are primarily of a preventative nature, which will assist in eliminating unexpected failure of a component or system, due to wear and tear, and increase safety margins when riding.

All the maintenance tasks are described in detail together with the procedures required for accomplishing them. If necessary, more general information on each topic can be found in the relevant Chapter within the main text.

Although no special tools are required for routine maintenance, a good selection of general workshop tools is essential. Included in the tools must be a range of metric ring or combination spanners and a selection of crosshead screwdrivers. Additionally, owing to the extreme tightness of most casing screws on Japanese machines, an impact screwdriver, together with a choice of large or small cross-head screw bits, is absolutely indispensable. This is particularly so if the engine has not been dismantled since leaving the factory.

---

**Weekly, or every 200 miles (300 km)**

---

## 1 Safety inspection

Give the complete machine a close and thorough visual inspection, checking for loose nuts, bolts and fittings, frayed control cables, damaged brake hoses, severe oil and petrol leaks, etc.

## 2 Legal inspection

Check the operation of the electrical system, ensuring that the lights and horn are working properly and that the lenses are clean. Note that in the UK it is an offence to use a vehicle on which the lights are defective. This applies even when the machine is used during daylight hours. The horn is also a statutory requirement.

## 3 Tyre pressures

Check the tyre pressures. Always check with the tyres cold, using a pressure gauge known to be accurate. It is recommended that a pocket pressure gauge is purchased to offset any fluctuation between garage forecourt instruments. The tyre pressures should be as follows:

|  | Solo | With pillion |
|---|---|---|
| *Front:* | | |
| GP100 | 25 psi (1.75 kg/cm²) | 25 psi (1.75 kg/cm²) |
| GP125 | 25 psi (1.75 kg/cm²) | 25 psi (1.75 kg/cm²) |
| *Rear:* | | |
| GP100 | 28 psi (2.00 kg/cm²) | 32 psi (2.25 kg/cm²) |
| GP125 | 32 psi (2.25 kg/cm²) | 36 psi (2.50 kg/cm²) |

At this juncture also inspect the actual condition of the tyres, ensuring there are no splits or cracks which may develop into serious problems. Also remove any small stones or other small objects of road debris which may be lodged between the tread blocks. A small flat-bladed screwdriver will be admirable for this job. Examine the amount of tread remaining on the tyre. The manufacturer's recommended minimum tread depth is 1.6 mm (0.06 in). A tyre with a tread depth below this figure should be renewed.

Check the tyre pressures with an accurate gauge

Check the tyre tread depth

## 4   Engine oil level

The oil tank level should be checked on a daily basis, prior to starting the engine. A small plastic window or sight glass gives an immediate visual warning of whether the oil level has dropped too low. Although it is quite safe to use the machine as long as oil is visible in the sight glass, it is recommended that the level is maintained to within about an inch of the tank filler neck, to allow a good reserve. It is advised that the oil tank level is checked whenever the machine is refuelled, and topped up as required.

## 5   Control cable lubrication

Apply a few drops of motor oil to the exposed inner portion of each control cable. This will prevent drying-up of the cables before the more thorough lubrication that should be carried out during the 2000 mile/3 monthly service.

## 6   Final drive chain lubrication

In order that the life of the final drive chain be extended as much as possible, regular lubrication is essential. Intermediate lubrication should take place with the chain in position on the machine. The chain should be lubricated by the application of one of the proprietary chain greases contained in an aerosol can. Ordinary engine oil can be used, though owing to the speed with which it is flung off the rotating chain, its effective life is limited.

Check the engine oil level through the sight glass

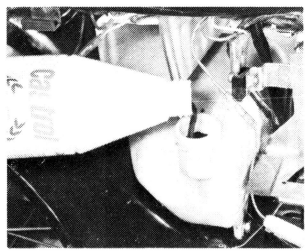

Keep the engine oil well topped up

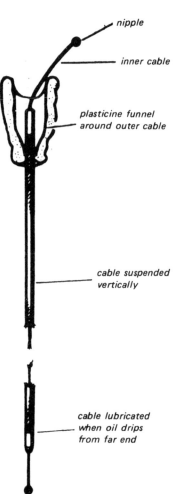

Control cable oiling

Use an aerosol chain lubricant at regular intervals

## 7 Hydraulic fluid level – disc brake models

Place the machine on its main stand on an area of flat and level ground. Position the handlebars in the central position so that the fluid reservoir is vertical. The fluid level can be seen through the translucent wall of the reservoir and should be between the upper and lower level marks. If necessary, remove the reservoir cap and replenish the reservoir with an hydraulic fluid of SAE J1703 specification. No other specification of fluid should be used, as an incorrect fluid may perish rubber seals within the brake system which will in turn cause brake failure. Note that hydraulic fluid will damage paintwork and plastic component parts and should, therefore, be wiped up as soon as a spillage occurs. If the fluid level in the reservoir is seen to be excessively low, then suspect a fluid leakage in the system.

Check the wheel alignment marks

Maintain the brake fluid level between the two level marks

Lock each adjuster bolt in position with the locknut

---

**Monthly, or every 600 miles (1000 km)**

---

Complete the tasks listed under the weekly/200 mile heading and then carry out the following checks:

## 1 Final drive chain adjustment

Check the slack in the final drive chain. The correct up and down movement, as measured at the mid-point of the chain lower run, should be 15 – 20 mm (0.6 – 0.8 in). Adjustment should be carried out as follows. Place the machine on the centre stand so that the rear wheel is clear of the ground and free to rotate. Remove the split pin from the wheel spindle and slacken the wheel nut. Loosen the locknuts on the two chain adjuster bolts, and slacken off the brake torque rod nuts.

Rotation of the adjuster bolts in a clockwise direction will tighten the chain. Tighten each bolt a similar number of turns so that wheel alignment is maintained. This can be verified by checking that the mark on the outer face of each chain adjuster is aligned with the same aligning mark on each fork end. With the adjustment correct, tighten the wheel nut to a torque setting of 3.6 - 5.2 kgf m (26.0 - 37.5 lbf ft) and lock it in position with a new split-pin. Spin the rear wheel to ensure that it rotates freely, adjust the rear brake operating mechanism (where necessary) and retighten the torque arm retaining nuts to a torque setting of 1.0 - 1.5 kgf m (7.0 - 11.0 lbf ft). Tighten the adjuster bolt locknuts.

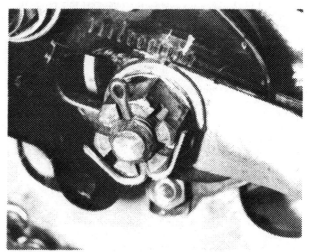

Lock the wheel spindle nut with a new split-pin

## 2  Battery electrolyte level

Check the battery electrolyte level. To gain access to the battery, remove the right-hand side panel by unscrewing the single retaining screw from its base and lifting it upwards off its two upper attachment points. If the electrolyte level does not lie between the upper and lower level marks on the battery case, then replenish each cell with distilled water until the level rises to the upper mark. Take care not to overfill the battery.

## Three monthly, or every 2000 miles (3000 km)

Complete the checks listed under the preceding Routine Maintenance Sections and then complete the following:

## 1  Final drive chain lubrication

Lubrication of the final drive chain should be carried out at short intervals as described in Section 6 of the weekly/200 mile maintenance schedule. A more thorough lubrication of the chain should be carried out every three months/2000 miles by carrying out the following procedure.

Separate the chain by removing the spring link and run it off the sprockets. If an old chain is available, interconnect the old and new chain, before the new chain is run off the sprockets. In this way the old chain can be pulled into place on the sprockets and then used to pull the regreased chain into place with ease.

Clean the chain thoroughly in a paraffin bath and then finally with petrol. The petrol will wash the paraffin out of the links and rollers which will then dry more quickly.

Allow the chain to dry and then immerse it in a molten lubricant such as Linklyfe or Chainguard. These lubricants must be used hot and will achieve better penetration of the links and rollers. They are less likely to be thrown off by centrifugal force when the chain is in motion.

Refit the newly greased chain onto the sprocket, replacing the spring link. This is accomplished most easily when the free ends of the chain are pushed into mesh on the rear wheel sprocket. The spring link must be fitted so that the closed end faces the normal direction of chain travel. Adjust the chain to the correct tension before the machine is used.

## 2  Control cable lubrication

Lubricate the control cables thoroughly with motor oil or an all-purpose oil. A good method of lubricating the cables is shown in the accompanying illustration, using a plasticine funnel. This method has the disadvantage that the cables usually need removing from the machine. An hydraulic cable oiler which pressurises the lubricant overcomes this problem. Do not lubricate nylon lined cables (which may have been fitted as replacements), as the oil may cause the nylon to swell, thereby causing total cable seizure.

## 3  General checks and lubrication

Work around the machine, applying grease or oil to any pivot points. These points should include the handlebar lever pivots, both stand pivots, the rear brake pedal pivot and the throttle twistgrip. Check all around the machine ensuring that all nuts, bolts and other fasteners are correctly secured. Check particularly that the engine mountings are fastened to the specified torque settings. Check the operation of the stand return springs; they should hold each stand securely in its retracted position.

## 4  Oil pump adjustment

In order to check the oil pump for correct adjustment, it is first necessary to remove both the rear section of the left-hand crankcase cover and the carburettor cover from the forward section of the right-hand crankcase cover. Remove the pump cover plate.

Check the pump for adjustment by rotating the throttle twistgrip until the circular indicator mark on the base of the

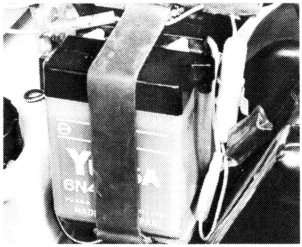

Remove the right-hand side panel to expose the battery

Fit the spring link correctly

Rotate the oil pump cable adjuster ...

throttle slide is seen to align with the upper edge of the carburettor mouth (see accompanying figure). With the throttle set in this position, check that the mark scribed on the pump lever boss is in exact alignment with the mark cast on the pump body. If this is not the case, then the marks should be made to align by rotating the control cable adjuster, after having first released its locknut. On completion of the adjustment procedure, retighten the locknut whilst holding the cable adjuster in position and then slide the rubber sealing cap back down the cable to cover the adjuster.

It should be noted that any adjustment of the oil pump control cable may well affect the adjustment of the throttle cable. It is, therefore, necessary to check the throttle cable for correct adjustment before starting the engine. Details of doing this are contained in this Chapter.

On completion of adjustment, refit all the disturbed covers. Check that the gasket of the carburettor cover is in a satisfac-

... to bring the pump lever mark into alignment with that on the pump body

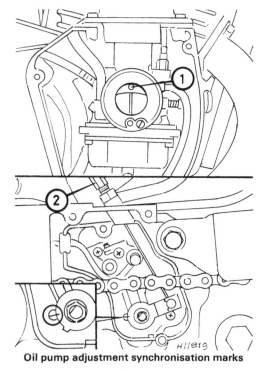

**Oil pump adjustment synchronisation marks**

*1   Carburettor punch mark     2   Cable adjuster*

tory condition before refitting it. Any air allowed past the gasket will cause the fuel/air mixture strength to weaken, thus adversely affecting the performance of the machine.

## 5   Changing the gearbox oil

In order to preclude any risk of the gearbox components having to run in oil that has deteriorated to the point where it is ineffective in its prime function as a lubricating medium, it is advisable to change the gearbox oil every three months/2000 miles.

Start the machine and run it until the engine reaches its normal operating temperature. Doing this will thin the oil in the gearbox thereby allowing it to drain more effectively. Stop the engine and position the machine on a flat and level piece of ground. Place the machine on its centre stand and position a container of at least 850 cc (1.50 Imp pint) capacity beneath the engine unit. Wipe clean the area of crankcase around the gearbox filler plug and the drain plug. Remove the drain plug, which is located on the underside of the right-hand crankcase half, and allow the oil to drain from the gearbox.

It will be seen that the drain plug is in fact the housing for the gearbox detent plunger and spring. Whilst waiting for the gearbox to drain, examine the detent assembly by following the procedure given in Section 4 of Chapter 1.

On completion of draining, refit the drain plug with its new washer and tighten it fully. Remove both the plastic filler plug and the oil level screw, which is located just forward of the kickstart shaft. Replenish the gearbox by pouring approximately

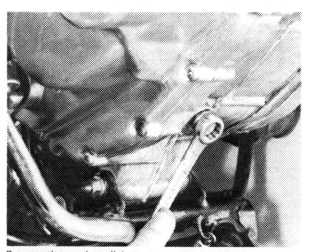

Remove the gearbox oil drain plug

Remove the oil level screw (arrowed) ...

... and replenish the gearbox with oil

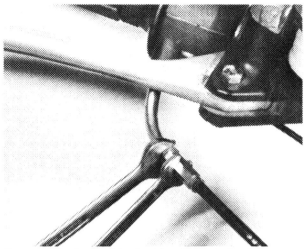

Carry out adjustment of the throttle cable at the handlebar adjuster

800 cc (1.41 Imp pint) of SAE 20W/40 oil through the filler orifice. The oil level is correct when oil begins to appear through the oil level hole. Refit the oil level screw with a new washer and tighten it. Refit and tighten the filler plug.

## 6   Carburettor adjustment

Adjustment of the carburettor should be carried out with the carburettor cover removed, the machine placed on its centre stand and the engine running at its normal operating temperature. The purpose of this adjustment is to ensure that the engine is running at its correct idle speed.

Once the engine has reached its normal operating temperature, stop it and remove the outermost of the two small blanking plugs from the front of the carburettor housing. The carburettor should now be adjusted so that the correct engine idle speed is obtained by first turning the pilot air screw fully in (until it seats lightly) and then unscrewing it $1\frac{1}{2}$ turns. Restart the engine and with the shank of the screwdriver inserted through the orifice in the carburettor housing, turn the throttle stop screw until the engine reaches its slowest possible reliable idle speed. The pilot air screw should now be turned slowly clockwise or anti-clockwise within a range of $\frac{1}{4}$ turn either side of its initial setting until the engine reaches the point where it idles most smoothly. At this point, the engine should be idling at the recommended speed of $1300 \pm 150$ rpm; if the reading on the tachometer indicates an idle speed which is slightly outside the recommended speed, then the throttle stop screw should be turned until the indicated speed is correct. Refit the small blanking plug and relocate the carburettor housing cover.

Always guard against the possibility of incorrect carburettor adjustments which will result in a weak mixture. Two-stroke engines are very susceptible to this type of fault, causing rapid overheating and often subsequent engine seizure. Changes in carburation leading to a weak mixture will occur if the air cleaner is removed or disconnected, or if the silencer is tampered with in any way.

## 7   Throttle cable adjustment

Adjustment of the throttle cable is correct when there is 1.0 - 1.5 mm (0.04 - 0.06 in) of free movement in the cable outer when it is pulled out of the adjuster at the handlebar end. If this adjustment is found to be incorrect, loosen the adjuster locknut and rotate the adjuster the required amount before holding it in position and retightening the locknut.

## 8   Cleaning the air filter element

The air filter assembly consists of a large oil-impregnated polyurethane foam filter element which is housed in a frame-mounted container located just to the rear of the cylinder barrel. This container is connected to the engine crankcase by means of a large-diameter rubber hose and attached to the frame by means of a mounting bracket with two screws and washers. The element itself must be removed from its container for the purposes of examination and cleaning.

To remove the element, move to the left-hand side of the machine and unscrew the single retaining bolt from the base of the housing cover. With the screw removed, the cover may be lifted up and away from the locating tab, which passes through its upper edge, and thus clear of the container. Press in on the element retaining plate and detach each of its retaining hooks. Remove the plate and withdraw the element.

Carry out a close inspection of the element. If the foam of the element shows signs of having become hardened with age or is seen to be very badly clogged, then it must be renewed. To clean the element, detach the foam from its metal frame and immerse it in a non-flammable solvent, such as white spirit, whilst gently squeezing it to remove any oil and dust. After cleaning, squeeze out the foam by pressing it between the palms of both hands and then allow a short time for any solvent remaining in the foam to evaporate. Do not wring out the foam as this will cause damage and thus lead to the need for early renewal.

Reimpregnate the foam with clean SAE 20W/40 oil and gently squeeze out any excess. Fit the metal frame into the foam so that the end of the frame is well inside the open end of the foam. Locate the element in the housing so that it fits correctly and secure its retaining plate in position with the two retaining hooks. Refit the cover to the end of the container and secure it in position with the single retaining bolt. Great care must be taken, when positioning both the element and the end cover, to ensure that no incoming air is allowed to bypass the element. If this is allowed to happen it will allow any dirt or dust that is normally retained by the element to find its way into the carburettor and crankcase assemblies; it will also effectively weaken the fuel/air mixture.

Note that if the machine is being run in a particularly dusty or moist atmosphere, then it is advisable to increase the frequency of cleaning and reimpregnating the element. Never run the engine without the element fitted. This is because the carburettor is specially jetted to compensate for the addition of this component and the resulting weak mixture will cause overheating of the engine with the probable risk of severe engine damage.

Remove the air filter housing cover ...

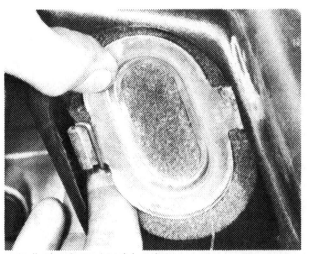

... unclip the element retaining plate ...

... and withdraw the air filter element

## 9 Checking the spark plug

Suzuki fit an NGK B8HS (or ND W24FS) spark plug as standard equipment to all the machines covered in this manual. The recommended gap between the plug electrodes is 0.6 - 0.8 mm (0.024 - 0.031 in). The plug should be cleaned and the gap checked and reset to regular service intervals. In addition, in the event of a roadside breakdown where the engine has mysteriously 'died', the spark plug should be the first item checked.

The plug should be cleaned thoroughly by using one of the following methods. The most efficient method of cleaning the electrodes is by using a bead blasting machine. It is quite possible that a local garage or motorcycle dealer has one of these machines installed on the premises and will be willing to clean any plugs for a nominal fee. Remember, before fitting a plug cleaned by this method, to ensure that there is none of the blasting medium left between the porcelain insulator and the plug body. An alternative method of cleaning the plug electrodes is to use a small brass-wire brush. Most motorcycle dealers sell such brushes which are designed specifically for this purpose. Any stubborn deposits of hard carbon may be removed by judicious scraping with a pocket knife. Take great care not to chip the porcelain insulator round the centre electrode whilst doing this. Ensure that the electrode faces are clean by passing a small fine file between them; alternatively, use emery paper but make sure that all traces of the abrasive material are removed from the plug on completion of cleaning.

To reset the gap between the plug electrodes, bend the outer electrode away from or closer to the central electrode and check that a feeler gauge of the correct size can be inserted between the electrodes. The gauge should be a light sliding fit.

Never bend the central electrode or the insulator will crack, causing engine damage if the particles fall in whilst the engine is running.

Always carry a spare spark plug fo the correct type. The plug in a two-stroke engine leads a particularly hard life and is liable to fail more readily than when fitted to a four-stroke.

Beware of overtightening the spark plug, otherwise there is risk of stripping the threads from the aluminium alloy cylinder head. The plug should be sufficiently tight to seat firmly on its sealing washer, and no more. Use a spanner which is a good fit to prevent the spanner from slipping and breaking the insulator.

Before fitting the spark plug in the cylinder head, coat its threads sparingly with a graphited grease. This will prevent the plug from becoming seized in the head and therefore aid future removal.

When reconnecting the suppressor cap to the plug, make sure that the cap is a good, firm fit and is in good condition;

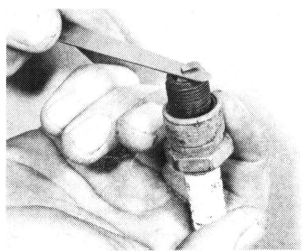

Check the spark plug electrode gap

renew its rubber seals if they are in any way damaged or
perished. The cap contains the suppressor that eliminates both
radio and TV interference.

### 10 Checking the contact breaker points

Access to the contact breaker assembly can be gained by
removing the inspection cover from the forward section of the
left-hand crankcase cover. This will reveal the flywheel generator rotor, which has four elongated holes in its outer face. The
largest two of these holes are provided to permit inspection and
adjustment of the contact breaker points.

Using the flat of a small screwdriver, open the contact
breaker points against the pressure of the return spring so that
the condition of the point contact faces may be checked. A
piece of stiff card or crokus paper may be used to remove any
light surface deposits, but if burnt or pitted, the complete
contact breaker assembly must be removed to facilitate further
examination of the point contact faces, and if necessary,
renewal of the aassembly. Refer to Section 4 of Chapter 3 for
full details on removal, renovation and fitting of the assembly. If
the points are found to be in sound condition, then proceed with
adjustment as follows.

Rotate the crankshaft slowly until the contact breaker
points are seen to be in their fully open position. Measure the
gap between the points with a feeler gauge. If the gap is
correct, a gauge of 0.35 mm (0.014 in) thickness will be a light
sliding fit between the point faces. If this is not the case, slacken
the single crosshead screw which serves to return the fixed
contact in position, just enough to allow movement of the
contact. With the flat of a screwdriver placed in the indentation
provided on the edge of the fixed contact plate, move the plate
in the appropriate direction until the gap is correct. Retighten
the retaining screw and recheck the gap setting; it is not
unknown for this setting to alter slightly upon retightening of
the screw.

Apply one or two drops of light machine oil to the cam
lubricating wick whilst taking care not to allow excess oil to foul
the point contact surfaces. Any adjustment of the contact
breakers must be followed by a check on the ignition timing as
detailed in the following Section.

### 11 Checking the ignition timing

It cannot be overstressed that optimum performance of the
engine depends on the accuracy with which the ignition timing
is set. Even a small error can cause a marked reduction in
performance and the possibility of engine damage as the result
of over-heating.

Static timing of the ignition can be carried out simply by
aligning the timing mark scribed on the wall of the flywheel
generator rotor with the corresponding mark on the crankcase
and then checking that the contact breaker points are just on
the point of separation. In order to gain access to the wall of the
rotor, it will be necessary to remove the forward section of the
left-hand crankcase cover.

Before commencing a check of the ignition timing, refer to
the previous Section of this Chapter and check that the contact
breaker points are clean and correctly gapped. In order to
provide an accurate indication as to when the contact breaker
points begin to separate, it will be necessary to obtain certain
items of electrical equipment. This equipment may take the
form of a multimeter or ohmmeter, or a high wattage 6 volt
bulb, complete with three lengths of electrical lead, which will
be used in conjunction with the battery of the motorcycle.

When carrying out the above described method of static
timing and the method of dynamic timing described at the end
of this Section, note that the accuracy of both methods of
timing depends very much on whether the flywheel generator
rotor is set correctly on the crankshaft. Any amount of wear
between the keyways in both the crankshaft and rotor bore and
the Woodruff key will cause some amount of variation between
the timing marks which will, in turn, lead to inaccurate timing.
Inaccuracy in the timing mark may also result from variations

Check the contact breaker gap

Reset the contact breaker gap by moving the fixed contact plate

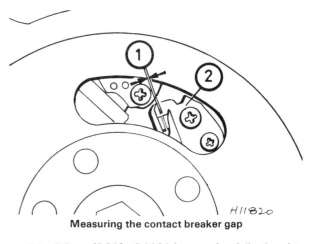

**Measuring the contact breaker gap**

1   0.3 - 0.4 mm (0.012 - 0.016 in)          2   Adjusting plate

during manufacture. The only means of overcoming this is to remove any movement between the two components and then to set the piston at a certain position within the cylinder bore before checking that the timing marks have remained in correct alignment. In order to position accurately the piston, it will be necessary to remove the spark plug and replace it with a dial gauge or a slide gauge, either one of which is adapted to fit into the spark plug hole of the cylinder head.

Position the piston in the cylinder bore by first rotating the crankshaft until the piston is set in the top dead centre (TDC) position. Set the gauge at zero on that position and then rotate the crankshaft backwards (clockwise) until the piston has passed down the cylinder bore a distance of at least 4 mm (0.16 in). Reverse the direction of rotation of the crankshaft until the piston is exactly 1.87 mm (0.074 in) from TDC. The timing marks should now be in exact alignment. If this is not the case, then new marks will have to be made. These can be used for all subsequent ignition timing checks.

Commence a check of the ignition timing by tracing the electrical lead from the fixed contact point (colour code, Black/yellow) to a point where it can be disconnected. To set up a multimeter, set it to its resistance function, connect one of its probes to the lead end and the other probe to a good earthing point on the crankcase. Use a similar method to set up an ohmmeter. In each case, opening of the contact breaker points will be indicated by a deflection of the instrument needle from one reading of resistance to another.

When using the battery and bulb method, remove the battery from the machine and position it at a convenient point next to the left-hand side of the engine. Connect one end of a wire to the positive (+) terminal of the battery and one end of another wire to the negative (–) terminal of the battery. The negative lead may now be earthed to a point on the engine casing. Ensure the earth point on the casing is clean and that the wire is positively connected, a crocodile clip fastened to the wire is ideal. Take the free end of the positive lead and connect it to the bulb. The third length of wire may now be connected between the bulb and the electrical lead of the fixed contact point. As the final connection is made, the bulb should light with the points closed. Opening of the contact breaker points will be indicated by a dimming of the bulb. The reason for recommending the use of a high wattage bulb is that this dimming will be more obvious to the eye.

Ignition timing is correct when the contact breaker points are seen to open just as the timing marks come into alignment. If this is not the case, then the points will have to be adjusted by moving the stator plate clockwise or anti-clockwise, depending on whether the opening point needs to be advanced or retarded. This is accomplished by removing the flywheel generator rotor in order to gain full access to the three stator plate retaining screws, each one of which passes through an elongated slot cut in the plate.

Removal of the flywheel rotor may be achieved by carrying out the following procedure. Prevent the crankshaft from rotating by selecting top gear and then applying the rear brake. Remove the rotor retaining nut, spring washer and plain washer. The rotor is a tapered fit on the crankshaft end and is located by a Woodruff key; it therefore requires pulling from position. The rotor boss is threaded internally to take the special Suzuki service tool No 09930 - 30102 with an attachment, No 09930 - 30161. If this tool cannot be acquired, then it is possible to remove the rotor by careful use of a two-legged puller.

If it is decided to make use of a two-legged puller, great care must be taken to ensure that both the threaded end of the crankshaft and the rotor itself remain free from damage. To protect the crankshaft end, refit the rotor retaining nut and screw it on until its outer face lies flush with the end of the crankshaft. Assemble the puller, checking that its feet are fitted through the two larger slots in the rotor face and are resting securely on the strengthened hub. Remember that the closer the feet of the puller are to the centre of the rotor then the less chance there is of the rotor becoming distorted. Gradually tighten the centre bolt of the puller to apply pressure to the rotor. Do not overtighten this bolt but apply reasonable pressure and then strike the end of the bolt with a hammer to break the taper joint. If this fails at first, tighten the centre bolt to apply a little more pressure and then try shocking the rotor free again.

It will be appreciated that resetting of the timing is a repetitive process of loosening the three stator plate retaining screws, rotating the stator plate a small amount in the required direction, retightening the screws, pushing the rotor back onto the crankshaft taper and then rechecking the timing. With the ignition timing correctly set, check tighten the stator plate retaining screws and then refit the rotor by first cleaning and degreasing both the taper of the crankshaft and the bore of the rotor where the two components come into contact. Check that the Woodruff key is correctly fitted in the crankshaft keyway and then push the rotor onto the crankshaft. Gently tap the centre of the rotor with a soft-faced hammer to seat it on the crankshaft taper and then fit the plain washer followed by the spring washer. Clean and degrease the threads of the rotor retaining nut and of the crankshaft end. Apply a thread locking compound to these threads and fit and tighten the nut, finger-tight. Lock the crankshaft in position by employing the method used for rotor removal and tighten the rotor retaining nut to a torque loading of 3.0 - 4.0 kgf m (21.5 - 29.0 lbf ft).

An alternative method of checking the ignition timing can be adopted, whilst the engine is running, using a stroboscopic lamp. This will entail gaining access to the wall of the rotor by removing the forward section of left-hand crankcase cover. When the light from the lamp is aimed at the timing marks on the crankcase and rotor wall, it has the effect of 'freezing' the moving mark on the rotor in one position and thus the accuracy of the timing can be seen.

Prepare the timing marks by degreasing them and then coating each one with a trace of white paint. This is not absolutely necesssary but will make the position of each mark far easier to observe if the light from the 'strobe' is weak or if the timing operation is carried out in bright conditions.

Two basic types of stroboscopic lamp are available, namely the neon and xenon tube types. Of the two, the neon type is much cheaper and will usually suffice if used in a shaded position, its light output being rather limited. The brighter but more expensive xenon types are preferable, if funds permit, because they produce a much clearer image.

Connect the 'strobe' to the HT lead, following the maker's instructions. If an external 6 volt power source is required, use the battery from the machine but make sure that when the leads from the machine's electrical system are disconnected from the terminals of the battery, they have their ends properly insulated with tape in order to prevent them from shorting on the cycle components.

Start the engine and aim the 'strobe' at the timing marks. The ignition timing is correct when the mark on the crankcase is in exact alignment with the mark on the rotor wall. To adjust the ignition timing, follow the instructions given in paragraphs 8, 9, 10 and 11 of this Section.

Finally, on completion of either one of the above mentioned timing procedures, refit the disturbed rotor cover and remove all test equipment from the machine.

## 12 Clutch adjustment

Check the amount of free play present in the clutch operating cable and the adjustment of the clutch release mechanism by following the instructions given below.

Remove the four screws that retain the outer cover to the forward section of the left-hand crankcase cover and pull the outer cover from position. Slide the rubber sealing cap up off the adjuster at the crankcase end of the cable. Commence adjustment by loosening the knurled lock ring of the handlebar lever adjuster and rotating the adjuster until it abuts against the lever bracket. Move to the adjuster at the crankcase end of the cable and loosen its locknut. Rotate the adjuster until free play can be felt in the cable inner; this allows correct adjustment of the operating mechanism. It is now necessary to free the adjuster screw contained within the end of the clutch operating

Commence clutch adjustment by rotating the operating arm adjuster screw

Set the clutch cable free play

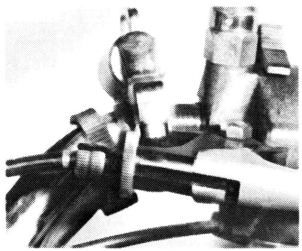

Fine adjustment can be achieved by rotating the knurled adjuster

arm by releasing the adjuster locknut. Turn this screw inwards until its end is felt to contact the clutch pushrod and then turn it outwards between $\frac{1}{4}$ and $\frac{1}{2}$ a turn. Lock the screw in position by retightening its locknut whilst taking care to ensure that the screw is not allowed to move from its set position. Complete the adjustment procedure by rotating the adjuster at the crankcase end of the cable to give 2 - 3 mm (0.08 - 0.12 in) of free play (measured between the pivot end of the handlebar lever and its retaining clamp) in the cable inner. Hold the adjuster in position and tighten its locknut. Any subsequent fine adjustment to the amount of free play in the cable inner may be achieved by rotating the knurled adjuster at the handlebar lever end before locking it in position by tightening its lock ring. Slide the rubber sealing cup into position over the adjuster at the crankcase end of the cable and refit the outer cover, complete with a serviceable gasket.

### 13 Check tightening the cylinder head nuts

Suzuki recommend that the nuts which retain the cylinder head in position and the two bolts which retain the exhaust pipe to the cylinder barrel be checked for security at regular service intervals. Any loosening of the cylinder head retaining nuts whilst the machine is in use will result in leakage around the head to cylinder barrel mating face, with the subsequent risk of distortion of the cylinder head casting. Leakage of exhaust gases from the pipe to cylinder barrel seal will affect the performance of the engine.

Check tighten the four cylinder head retaining nuts by torque loading them to 2.3 - 2.7 kgf m (16.5 - 19.5 lbf ft). Tighten in even increments whilst working in a diagonal sequence. If the nuts are found to be loose and the cylinder head to barrel joint leaking, then the cylinder head gasket must be renewed.

If the exhaust pipe retaining bolts are found to be loose, then remove them and check the condition of the spring washers fitted beneath their heads. If these washers have become flattened then they must be renewed before the bolts are refitted and tightened. It is a good idea to renew the sealing ring between the barrel and pipe to prevent leakage.

### 14 Checking front brake pad wear – disc brake models

The degree of wear present on each brake pad can be checked by viewing the pads from the front of the machine. Each pad is inscribed with a red groove which marks the wear limit of the friction material. When this limit is reached, both pads must be renewed, even if only one has reached the wear mark. In normal use, both pads will wear at the same rate and both must therefore be renewed.

Before attempting renewal of the brake pads, thoroughly clean the area of machine around the brake caliper. This will prevent any ingress of road dirt into the caliper assembly. To gain access to the pads for renewal, detach the caliper assembly from the fork leg by removing the two securing bolts, each with its plain washer and spring washer. Disconnection of the hydraulic hose is not required.

Remove the single screw and the convolute backing plate from the inner side of the caliper unit. The inner pad is now free and may be displaced towards the centre of the caliper and lifted out. The outer pad which abuts against the caliper piston is not retained positively and may be lifted out.

Fit the new pads and refit the caliper assembly to the fork leg by reversing the dismantling procedure, whilst noting the following points. Take care not to operate inadvertently the brake lever at any time whilst the caliper assembly is removed from the brake disc. Doing this will displace the caliper piston, thereby making reassembly considerably more difficult. Before fitting the new pads, push the caliper piston inwards slightly so that there is sufficient clearance between the brake pads to allow the caliper to fit over the disc. It is recommended that the outer periphery of the outer (piston) pad is lightly coated with disc brake assembly grease (silicon grease). Use the grease

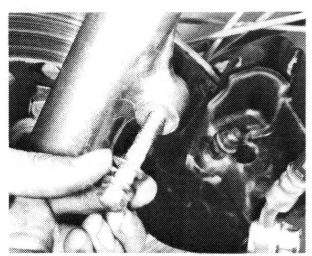

Remove the brake caliper securing bolts

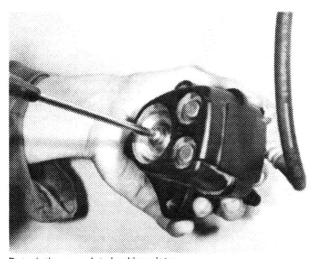

Detach the convolute backing plate ...

... and withdraw the inner brake pad

sparingly and ensure that grease **DOES NOT** come in contact with the friction surface of the pad.

Before refitting the caliper to the fork leg securing bolts, check that the spring washer located beneath each bolt head has not become flattened. If this is the case, then the washer must be renewed as it will no longer serve its locking function. Tighten each of these bolts to a torque loading of 1.5 - 2.5 kgf m (11.0 - 18.0 lbf ft).

Finally, in the interests of safety, always check the function of the brakes; pump the brake lever several times to restore full braking power, before taking the machine on the road.

### 15 Front brake adjustment – drum brake models

Adjustment of the front drum brake is correct when there is 20 - 30 mm (0.8 - 1.2 in) of clearance between the end of the handlebar lever and the throttle twistgrip with the lever fully applied.

To adjust the clearance between the lever and twistgrip, simply turn the nut at the camshaft operating arm end of the cable the required amount in the appropriate direction. Any minor adjustments necessary may then be made with the cable adjuster at the handlebar lever bracket. To use this adjuster, simply loosen the lock ring, turn the knurled adjuster the required amount and then retighten the lock ring.

On completion of adjustment, check the brake for correct operation by spinning the wheel and applying the brake lever. There should be no indication of the brake binding as the wheel is spun. If the brake shoes are heard to be brushing against the surface of the wheel drum, back off on the cable adjuster slightly until all indication of binding disappears. The brake may be readjusted after a period of bedding-in has been allowed for the brake shoes.

### 16 Rear brake adjustment

adjustment of the rear brake is correct when there is 20 - 30 mm (0.8 - 1.2 in) of movement, measured at the forward point of the brake pedal, between the point at which the brake pedal is fully depressed and the point where it abuts against its return stop.

The range of pedal movement may be adjusted simply by turning the nut on the wheel end of the brake operating rod in the required direction.

On completion of brake adjustment, check that the stop lamp switch operates the stop lamp as soon as the brake pedal is depressed. If necessary, adjust the height setting of the switch in accordance with the instructions given in Chapter 6.

### 17 Checking brake shoe wear – drum brakes

An indication of brake shoe lining wear is provided by an

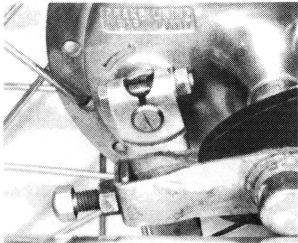

An indication of brake shoe lining wear is provided

indicator line which is cast into the brake backplate. If, when the brake is correctly adjusted and applied fully, the line on the end of the brake cam spindle is seen to align with a point outside the arc shown by the indicator line, then the lining on the brake shoes can be assumed to have worn beyond limits and should be renewed at the earliest possible opportunity.

---

### Six monthly, or every 4000 miles (6000 km)

Complete the checks listed under the preceding Routine Maintenance Sections and then complete the following:

### 1   Spark plug renewal

Remove and discard the existing spark plug, regardless of its condition. Although the plug may give acceptable performance after it has reached this mileage, it is unlikely that it is still working at peak efficiency.

The correct type of spark plug is an NGK B8HS or an ND W24FS. Before fitting the new plug, adjust the gap between the electrodes to 0.6 - 0.8 mm (0.024 - 0.031 in), coat its threads sparingly with graphite grease and check that the aluminium crush washer is in place on the plug.

### 2   Cleaning the fuel tap filter element

The type of fuel tap fitted to the machines covered in this Manual incorporates a fuel filter element which is located within the base of the tap and retained in position by a plastic bowl and O-ring assembly. The purpose of this filter element is to prevent any contamination in the fuel from being passed directly from the tank to the carburettor, thus precluding the likelihood of any of the jets within the carburettor becoming blocked. Contamination caught by the filter is deposited in the plastic bowl which must threrefore be removed, together with the filter element, at frequent intervals for cleaning and examination. The recommended service interval for these components is every 4000 miles (6000 km). It should however, be realised that any loss in engine performance or a refusal of the engine to run for any more than a short period of time, might be attributable to fuel starvation caused by a blocked or partially blocked filter element. Any suspected contamination of the fuel will therefore lead to the need to clean the element and bowl at more frequent intervals than recommended. It may well also be necessary to remove and clean out the fuel tank.

To remove the filter element turn the top lever to the 'Off' position, place an open-ended spanner over the squared end of

the bowl and turn it anti-clockwise to unscrew the bowl. It was found in practice that the O-ring between the bowl and the tap casing had formed a semi-permanent seal between the two components and some effort was required to effect an initial release of the bowl. With the bowl and O-ring removed, the filter element may be gently eased out of its location within the tap whilst taking care to note the positioning of the hole within the filter element in relation to the corresponding fuel line within the tap.

Inspect the condition of the O-ring; if it is flattened, perished, or in any way damaged, then it must be renewed. Clean the filter element by rinsing it in clean fuel. Any stubborn traces of contamination may be removed from the element by gently brushing it with a small soft-bristled brush soaked in fuel, a used toothbrush is ideal. The filter bowl may be cleaned by using a similar method. Remember to take the necessary fire precautions when carrying out these cleaning procedures and always wear eye protection against any fuel that may spray back from the brush. Once it is cleaned, closely inspect the gauze of the element for any splits or holes that will allow the passage of sediment through it and onto the carburettor. Renew the element if it is in any way defective.

Clean the fuel tap base of any remaining sediment before inserting the element into position, fitting the serviceable O-ring and fitting and tightening the filter bowl. Take care not to overtighten the bowl, it need only be nipped tight. Finally turn the tap lever to the 'On' position and carry out a check for any leakage of fuel around the bowl to tap joint. Cure any leak found by nipping the bowl a little tighter. If this fails, remove the bowl and check that the O-ring has seated correctly.

### 3   Decarbonising the cylinder head and barrel

Decarbonising of the cylinder head and cylinder barrel can be undertaken with a minimal amount of dismantling, namely removal of the cylinder head, cylinder barrel and, as an additional requirement, the exhaust silencer baffle.

Carbon build up in a two-stroke engine is more rapid than that of its four-stroke counterpart, due to the oily nature of the combustion mixtue. It is however, rather softer and is therefore more easily removed.

The object of the exercise is to remove all traces of carbon whilst avoiding the removal of the metal surface on which it is deposited. It follows that care must be taken when dealing with the relatively soft alloy cylinder head and piston. Never use a steel scraper or screwdriver for carbon removal. A hardwood, brass or aluminium scraper is the ideal tool as these are harder than the carbon, but no harder than the underlying metal. Once the bulk of the carbon has been removed, a brass wire brush of

Remove the bowl from the fuel tap ...

... and inspect the fuel filter component parts

the type used to clean suede shoes can be used to good effect.

The whole of the combustion chamber should be cleaned, as should the piston crown. It is recommended that as smooth a finish as possible is obtained, as this will slow the subsequent build up of carbon. If desired, metal polish can be used to obtain a smooth surface. The exhaust port must also be cleaned out, as a build up of carbon in this area will restrict the flow of exhaust gases from the cylinder. Take care to remove all traces of debris from the cylinder and ports, prior to reassembly, by washing the components thoroughly in fuel or paraffin whilst taking care to observe the necessary fire precautions.

Full details of decarbonising the silencer assembly are given in Section 16 of Chapter 2.

### 4 Steering head bearing check and adjustment

Place the machine on the centre stand so that the front wheel is clear of the ground. If necessary, place blocks below the crankcase to prevent the motorcycle from tipping forwards.

Grasp the front fork legs near the wheel spindle and push and pull firmly in a fore and aft direction. If play is evident between the upper and lower steering yokes and the head lug casting, the steering head bearings are in need of adjustment. Imprecise handling or a tendency for the front forks to judder may be caused by this fault.

Bearing adjustment is correct when the adjuster ring is tightened until resistance to movement is felt and then loosened $\frac{1}{8}$ to $\frac{1}{4}$ of a turn. The adjuster ring should be rotated by means of a C-spanner.

Take great care not to overtighten the nut. It is possible to place a pressure of several tons on the head bearings by overtightening even though the handlebars may seem to turn quite freely. Overtight bearings will cause the machine to roll at low speeds and give imprecise steering. Adjustment is correct if there is no play in the bearings and the handlebars swing to full lock either side when the machine is on the centre stand with the front wheel clear of the ground. Only a light tap on each end should cause the handlebars to swing.

### 5 Speedometer and tachometer cable lubrication

To grease either the speedometer or tachometer cable, uncouple both ends and withdraw the inner cable. (On some model types this may not be possible in which case a badly seized cable will have to be renewed as a complete assembly). After removing any old grease, clean the inner cable with a petrol soaked rag and examine the cable for broken strands or other damage. Do not check the cable for broken strands by passing it through the fingers or palm of the hand, this may well cause a painful injury if a broken strand snags the skin. It is best to wrap a piece of rag around the cable and pull the cable through it, any broken strands will snag on the rag.

Regrease the cable with high melting point grease, taking care not to grease the last six inches closest to the instrument head. If this precaution is not observed, grease will work into the instrument and immobilise the sensitive movement.

### 6 Brake cam shaft lubrication – drum brakes

In order to gain access to the brake cam shaft of each drum brake assembly it is necessary to remove the wheel and withdraw the brake backplate from the wheel hub. It is a good idea to combine this operation with the examination sequence given in Section 14 of Chapter 5. Note that failure to lubricate the cam shaft could well result in seizure of the shaft during operation of the brake with disastrous consequences. After displacing the brake shoes the cam shaft can be removed by withdrawing the retaining bolt on the operating arm and pulling the arm off the shaft. Before removing the arm, it is advisable to mark its position in relation to the shaft, so that it can be relocated correctly.

Remove any deposits of hardened grease or corrosion from the bearing surface of the brake cam shaft by rubbing it lightly with a strip of fine emery paper or by applying solvent with a piece of rag. Lightly grease the length of the shaft and the face of the operating cam prior to reassembly. Clean and grease and pivot stub which is set in the backplate.

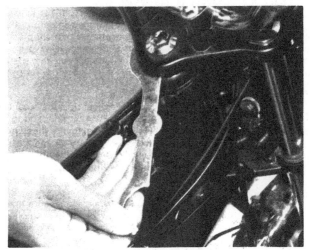
Do not overtighten the steering head bearings

Clean and lubricate the brake cam shaft

Align the brake operating arm with the cam shaft

Check the condition of the O-ring which prevents the escape of grease from the end of the camshaft. If it is in any way damaged or perished, then it must be renewed before the shaft is relocated in the backplate. Relocate the cam shaft and align and fit the operating arm with the O-ring and plain washer. The bolt and nut retaining the arm in position on the shaft should be torque loaded to 0.5 − 0.8 kgf in (3.5 − 6.0 lbf ft).

### 7   Check the wheels

Referring to Chapter 5.2 check the wheel rims for runout, the spokes for straightness, security and even tension, and the bearings for signs of free play. Any faults found must be rectified immediately. Also check the tyre pressures and tread depth as described under the weekly check; remember that excessively worn tyres must be renewed immediately.

### 8   Check the suspension

Examine closely the front and rear suspension. Ensure that the front forks work smoothly and progressively by pumping them up and down whilst the front brake is held on. Any faults revealed by this check should be investigated further. Check carefully for signs of leaks around the front fork oil seals. If any damage is found, it must be repaired immediately as described in the relevant Section of Chapter 4.

To check the swinging arm place the machine on its centre stand then pull and push horizontally at the rear end of the swinging arm; there should be no discernible play at the pivot.

---

**Annually, or every 8000 miles (12 000 km)**

---

Complete the checks listed under the preceding Routine Maintenance Sections and then complete the following:

### 1   Cleaning the carburettor

Suzuki recommend that the carburettor is removed and dismantled for cleaning at every 8000 mile service interval. This should only be necessary if there is real evidence of fuel contamination or if the performance of the machine is thought to be affected by a fault in carburation.

Evidence of fuel contamination will be found in the fuel tap filter element. Full details of servicing the carburettor may be found in Chapter 2 of this Manual.

### 2   Renewing the contact breaker points

Suzuki recommend that the contact breaker point assembly is renewed at every 8000 mile service interval. This is, of course, only necessary if the points have a substantial amount of material missing from their faces, due to routine dressing or spark erosion.

### 3   Change the front fork oil

This is an important task which must be carried out to ensure the continuing stability and safety of the machine on the road. Fork oil gradually degenerates as it loses viscosity and becomes contaminated by water and dirt, which produces a very gradual loss of damping. This can occur over a long period of time, thus being completely unnoticed by the rider until the machine is in a dangerous condition. Regular changes of the fork oil will eliminate this possibility. Refer to Chapter 4, Sections 2 and 3 for details of removal and refitting. Note that as most machines are not fitted with front fork drain plugs, each fork leg must be removed from the machine and inverted to drain the oil. This does, however, offer an excellent opportunity to examine the forks for wear and damage and fits in well with the other maintenance operations that are necessary at this interval.

### 4   Grease the wheel bearings and speedometer drive

While the wheels are removed from the machine during the course of the previous maintenance operation, take the opportunity to pack some grease into the speedometer drive, as described in Chapter 4, Section 18 and Chapter 5, Section 4. Check that the wheel bearings are in good condition and well lubricated, as described in Chapter 5, Section 5.

### 5   Renew the brake fluid − disc brake models

If the brake fluid is not completely changed during the course of routine maintenance, it should be changed at least once a year. Brake fluid is hygroscopic, which means that it absorbs moisture from the air. Although the system is sealed, the fluid will gradually deteriorate and must be renewed before contamination lowers its boiling point to an unsafe level.

Before starting work, obtain a full can of new DOT 3 or 4 or SAE J1703 hydraulic fluid and read Chapter 5, Section 11. Prepare the clear plastic tube and glass in the same way as for bleeding the hydraulic system, open the bleed nipple by unscrewing it $\frac{1}{4} - \frac{1}{2}$ a turn with a spanner and apply the front brake lever gently and repeatedly. This will pump out the old fluid. Keep the master cylinder reservoir topped up at all times, otherwise air may enter the system and greatly lengthen the operation. The old brake fluid is invariably much darker in colour than the new, making it much easier to see when the old fluid has been pumped out and the new fluid has completely replaced it.

When the new fluid appears in the clear plastic tubing with no traces of old fluid contaminating it, close the bleed nipple, remove the plastic tubing and refit the rubber dust cap on the nipple. Top the master cylinder reservoir up to the 'Upper' level mark. Carefully dry the diaphragm with a clean lint-free cloth, fold it into its compressed state, and refit the diaphragm and the reservoir cover, tightening securely the retaining screw. Wash off any surplus fluid with fresh water and check for any fluid leaks which may subsequently appear. Remember to check that full brake pressure is restored and that the front brake is working properly before taking the machine out on the road.

---

**Two yearly, or every 16 000 miles (24 000 km)**

---

### 1   Renewing the fuel feed pipe

Because the fuel feed pipe from the fuel tank to the carburettor is constructed of thin-walled synthetic rubber which will be affected by heat and the elements over a period of time, Suzuki recommend that, to anticipate any risk of fuel leakage, the pipe is renewed at every 16 000 mile service interval.

### 2   Renewing the brake hose − disc brake models

Suzuki recommend that in the interest of safety, the hydraulic hose of the front disc brake should be renewed at every 16 000 mile service interval. This is because constant contact of the hose with road salts, moisture, etc, will cause the hose material to harden and eventually split, which in turn will greatly increase the risk of fluid leakage.

### 3   Grease the steering head and swinging arm pivot bearings

If the steering head bearings have not been dismantled for any other reason, they should be stripped for examination and greasing every two years or 16 000 miles (24 000 km). This task would fit in conveniently with the fork oil change and is described in Chapter 4, Sections 2, 4 and 5.

Similarly the swinging arm should be removed for the pivot bearings to be examined and greased, if this has not been done previously.

# Conversion factors

## Length (distance)

| | | | | | |
|---|---|---|---|---|---|
| Inches (in) | X | 25.4 | = Millimetres (mm) | X 0.0394 | = Inches (in) |
| Feet (ft) | X | 0.305 | = Metres (m) | X 3.281 | = Feet (ft) |
| Miles | X | 1.609 | = Kilometres (km) | X 0.621 | = Miles |

## Volume (capacity)

| | | | | | |
|---|---|---|---|---|---|
| Cubic inches (cu in; in³) | X | 16.387 | = Cubic centimetres (cc; cm³) | X 0.061 | = Cubic inches (cu in; in³) |
| Imperial pints (Imp pt) | X | 0.568 | = Litres (l) | X 1.76 | = Imperial pints (Imp pt) |
| Imperial quarts (Imp qt) | X | 1.137 | = Litres (l) | X 0.88 | = Imperial quarts (Imp qt) |
| Imperial quarts (Imp qt) | X | 1.201 | = US quarts (US qt) | X 0.833 | = Imperial quarts (Imp qt) |
| US quarts (US qt) | X | 0.946 | = Litres (l) | X 1.057 | = US quarts (US qt) |
| Imperial gallons (Imp gal) | X | 4.546 | = Litres (l) | X 0.22 | = Imperial gallons (Imp gal) |
| Imperial gallons (Imp gal) | X | 1.201 | = US gallons (US gal) | X 0.833 | = Imperial gallons (Imp gal) |
| US gallons (US gal) | X | 3.785 | = Litres (l) | X 0.264 | = US gallons (US gal) |

## Mass (weight)

| | | | | | |
|---|---|---|---|---|---|
| Ounces (oz) | X | 28.35 | = Grams (g) | X 0.035 | = Ounces (oz) |
| Pounds (lb) | X | 0.454 | = Kilograms (kg) | X 2.205 | = Pounds (lb) |

## Force

| | | | | | |
|---|---|---|---|---|---|
| Ounces-force (ozf; oz) | X | 0.278 | = Newtons (N) | X 3.6 | = Ounces-force (ozf; oz) |
| Pounds-force (lbf; lb) | X | 4.448 | = Newtons (N) | X 0.225 | = Pounds-force (lbf; lb) |
| Newtons (N) | X | 0.1 | = Kilograms-force (kgf; kg) | X 9.81 | = Newtons (N) |

## Pressure

| | | | | | |
|---|---|---|---|---|---|
| Pounds-force per square inch (psi; lbf/in²; lb/in²) | X | 0.070 | = Kilograms-force per square centimetre (kgf/cm²; kg/cm²) | X 14.223 | = Pounds-force per square inch (psi; lbf/in²; lb/in²) |
| Pounds-force per square inch (psi; lbf/in²; lb/in²) | X | 0.068 | = Atmospheres (atm) | X 14.696 | = Pounds-force per square inch (psi; lbf/in²; lb/in²) |
| Pounds-force per square inch (psi; lbf/in²; lb/in²) | X | 0.069 | = Bars | X 14.5 | = Pounds-force per square inch (psi; lbf/in²; lb/in²) |
| Pounds-force per square inch (psi; lbf/in²; lb/in²) | X | 6.895 | = Kilopascals (kPa) | X 0.145 | = Pounds-force per square inch (psi; lbf/in²; lb/in²) |
| Kilopascals (kPa) | X | 0.01 | = Kilograms-force per square centimetre (kgf/cm²; kg/cm²) | X 98.1 | = Kilopascals (kPa) |
| Millibar (mbar) | X | 100 | = Pascals (Pa) | X 0.01 | = Millibar (mbar) |
| Millibar (mbar) | X | 0.0145 | = Pounds-force per square inch (psi; lbf/in²; lb/in²) | X 68.947 | = Millibar (mbar) |
| Millibar (mbar) | X | 0.75 | = Millimetres of mercury (mmHg) | X 1.333 | = Millibar (mbar) |
| Millibar (mbar) | X | 0.401 | = Inches of water (inH₂O) | X 2.491 | = Millibar (mbar) |
| Millimetres of mercury (mmHg) | X | 0.535 | = Inches of water (inH₂O) | X 1.868 | = Millimetres of mercury (mmHg) |
| Inches of water (inH₂O) | X | 0.036 | = Pounds-force per square inch (psi; lbf/in²; lb/in²) | X 27.68 | = Inches of water (inH₂O) |

## Torque (moment of force)

| | | | | | |
|---|---|---|---|---|---|
| Pounds-force inches (lbf in; lb in) | X | 1.152 | = Kilograms-force centimetre (kgf cm; kg cm) | X 0.868 | = Pounds-force inches (lbf in; lb in) |
| Pounds-force inches (lbf in; lb in) | X | 0.113 | = Newton metres (Nm) | X 8.85 | = Pounds-force inches (lbf in; lb in) |
| Pounds-force inches (lbf in; lb in) | X | 0.083 | = Pounds-force feet (lbf ft; lb ft) | X 12 | = Pounds-force inches (lbf in; lb in) |
| Pounds-force feet (lbf ft; lb ft) | X | 0.138 | = Kilograms-force metres (kgf m; kg m) | X 7.233 | = Pounds-force feet (lbf ft; lb ft) |
| Pounds-force feet (lbf ft; lb ft) | X | 1.356 | = Newton metres (Nm) | X 0.738 | = Pounds-force feet (lbf ft; lb ft) |
| Newton metres (Nm) | X | 0.102 | = Kilograms-force metres (kgf m; kg m) | X 9.804 | = Newton metres (Nm) |

## Power

| | | | | | |
|---|---|---|---|---|---|
| Horsepower (hp) | X | 745.7 | = Watts (W) | X 0.0013 | = Horsepower (hp) |

## Velocity (speed)

| | | | | | |
|---|---|---|---|---|---|
| Miles per hour (miles/hr; mph) | X | 1.609 | = Kilometres per hour (km/hr; kph) | X 0.621 | = Miles per hour (miles/hr; mph) |

## Fuel consumption*

| | | | | | |
|---|---|---|---|---|---|
| Miles per gallon, Imperial (mpg) | X | 0.354 | = Kilometres per litre (km/l) | X 2.825 | = Miles per gallon, Imperial (mpg) |
| Miles per gallon, US (mpg) | X | 0.425 | = Kilometres per litre (km/l) | X 2.352 | = Miles per gallon, US (mpg) |

## Temperature

Degrees Fahrenheit = (°C x 1.8) + 32        Degrees Celsius (Degrees Centigrade; °C) = (°F - 32) x 0.56

*It is common practice to convert from miles per gallon (mpg) to litres/100 kilometres (l/100km), where mpg (Imperial) x l/100 km = 282 and mpg (US) x l/100 km = 235

# Chapter 1 Engine, clutch and gearbox

## Contents

## Specifications

| Model | GP100 | GP125 |
|---|---|---|
| **Engine** | | |
| Type ................................................................................ | Air cooled, single cylinder, rotary disc valve, two-stroke | |
| Bore ................................................................................ | 50 mm (1.969 in) | 56 mm (2.205 in) |
| Stroke ............................................................................. | 50 mm (1.969 in) | 50 mm (1.969 in) |
| Capacity ......................................................................... | 98 cc (6.0 cu in) | 123 cc (7.5 cu in) |
| Compression ratio .......................................................... | 6.8 : 1 | 6.7 : 1 |
| Power output ................................................................. | 12 bhp @ 8500 rpm | 15 bhp @ 8500 rpm |
| **Cylinder barrel** | | |
| Standard bore ................................................................ | 50.0 – 50.015 mm (1.9685 – 1.9691 in) | 56.0 – 56.015 mm 2.2047 – 2.2053 in) |
| Service limit ................................................................... | 50.095 mm (1.9722 in) | 56.095 mm (2.2085 in) |
| Barrel to head face distortion limit ................................ | 0.05 mm (0.002 in) | |
| Cylinder bore to piston clearance ................................. | 0.035 – 0.045 mm (0.0014 – 0.0018 in) | |
| Service limit ................................................................... | 0.120 mm (0.0047 in) | |

### Cylinder head
Head to barrel face distortion limit ........................................... 0.05 mm (0.002 in)

### Piston
Outside diameter ................................................................ 49.960 – 49.975 mm      55.960 – 55.975 mm
(1.9669 – 1.9675 in)      (2.2013 – 2.2037 in)

Service limit ..................................................................... 49.880 mm      55.880 mm
(1.9638 in)      (2.20 in)

Gudgeon pin OD .................................................................. 13.995 – 14.0 mm (0.5510 – 0.5512 in)
Service limit ..................................................................... 13.980 mm (0.5504 in)
Gudgeon pin hole bore diameter .............................................. 13.998 – 14.006 mm (0.5511 – 0.5514 in)
Gudgeon pin to piston clearance ............................................. 0.002 – 0.011 mm (0.0001 – 0.0004 in)
Service limit ..................................................................... 0.080 mm (0.0031 in)

### Piston rings (top and second)
Ring to groove clearance ...................................................... 0.03 – 0.07 mm (0.001 – 0.003 in)
End gap:
Fitted ........................................................................ 0.15 – 0.35 mm (0.006 – 0.014 in)
Service limit ............................................................... 0.80 mm (0.031 in)
Free ......................................................................... 5.0 mm (0.20 in)      7.5 mm (0.30 in)
Nippon (N)
4.5 mm (0.18 in)
Teikoku (T)
Service limit ............................................................... 4.0 mm (0.16 in) N      6.0 mm (0.24 in)
3.6 mm (0.14 in) T

### Crankshaft assembly
Maximum runout ................................................................. 0.05 mm (0.002 in)
Maximum connecting rod deflection .......................................... 3.0 mm (0.12 in)
Maximum big-end radial play .................................................. 0.08 mm (0.003 in)
Small-end bore .................................................................. 18.003 – 18.011 mm (0.7088 – 0.7091 in)
Service limit ..................................................................... 18.040 mm (0.7102 in)

### Clutch
Type ............................................................................... Wet, multiplate
Spring free length .............................................................. 32.0 mm (1.26 in)
Service limit ..................................................................... 33.6 mm (1.32 in)
Friction plate:
Thickness .................................................................... 2.9 – 3.0 mm (0.11 – 0.12 in)
Service limit ............................................................... 2.6 mm (0.10 in)
Tongue width ............................................................... 11.8 – 12.0 mm (0.46 – 0.47 in)
Service limit ............................................................... 11.3 mm (0.45 in)
Plain plate:
Thickness .................................................................... 1.6 mm (0.063 in)
Maximum warpage ......................................................... 0.1 mm (0.004 in)

### Gearbox
Type ............................................................................... 5-speed, constant mesh
Gear ratios (no of teeth):
1st ........................................................................... 3.090 : 1 (34/11)
2nd .......................................................................... 1.812 : 1 (29/16)
3rd ........................................................................... 1.250 : 1 (25/20)
4th ........................................................................... 0.956 : 1 (22/23)
5th ........................................................................... 0.800 : 1 (20/25)
Backlash in all gears .......................................................... 0.10 mm (0.004 in)
Service limit ..................................................................... 0.15 mm (0.006 in)
Final reduction ratio:
GP100 UN, UX and early UD .............................................. 3.067:1 (46/15)
GP100 UL and later UD .................................................... 2.800:1 (42/15)
GP100 C models ............................................................ 3.285:1 (46/14)
GP100 N, X and D models ................................................. 3.067:1 (46/15)
GP125 models ............................................................... 2.733:1 (41/15) or 3.000:1 (45/15)
Primary reduction ratio ....................................................... 3.625 : 1 (58/16)
Primary drive gear backlash .................................................. 0.02 – 0.07 mm (0.001 – 0.003 in)
Service limit ..................................................................... 0.10 mm (0.004 in)
Selector fork claw end thickness:
No 1 .......................................................................... 4.30 – 4.40 mm (0.169 – 0.173 in)
No 2 .......................................................................... 5.30 – 5.40 mm (0.209 – 0.213 in)
Selector fork to pinion groove clearance:
Nos 1 and 2 ................................................................ 0.05 – 0.25 mm (0.002 – 0.010 in)
Service limit ............................................................... 0.45 mm (0.018 in)
Pinion groove width:
No 1 .......................................................................... 4.45 – 4.55 mm (0.175 – 0.179 in)
No 2 .......................................................................... 5.45 – 5.55 mm (0.215 – 0.219 in)

**Torque wrench settings**

| Component | kgf m | lbf ft |
|---|---|---|
| Cylinder head retaining nuts ........................................ | 2.3 – 2.7 | 16.5 – 19.5 |
| Primary drive pinion retaining nut ........................................... | 3.6 – 5.0 | 26.0 – 36.0 |
| Flywheel generator rotor retaining nut .................................... | 3.0 – 4.0 | 21.5 – 29.0 |
| Clutch hub retaining nut ........................................ | 2.0 – 3.0 | 14.5 – 21.5 |
| Gearbox sprocket retaining nut ................................................. | 4.0 – 6.0 | 29.0 – 43.5 |
| Engine mounting nuts: | | |
| 8 mm ...................................................................... | 1.8 – 2.8 | 13.0 – 20.2 |
| 10 mm ...................................................................... | 4.0 – 6.5 | 29.0 – 47.0 |

## 1  General description

The Suzuki GP100 and GP125 models covered in this Manual are all fitted with the same design of engine/gearbox unit, this being a single cylinder, air cooled two-stroke engine with a five-speed, constant mesh gearbox built in unit. Drive is transmitted from the crankshaft to the gearbox mainshaft via helical primary gears and a wet multiplate clutch.

Engine lubrication is effected by means of the Suzuki CCI system, in which oil is drawn from a frame-mounted tank and pressure-fed to the main and big-end bearings, and to the disc valve assembly. Other moving parts are lubricated by the residual oil. The system is of the constant-loss type, the residual oil being burnt and expelled with the exhaust gases. Metering of the supply is controlled by engine speed and throttle opening. Lubrication for the gearbox and primary transmission is provided by an oil reservoir shared between the two interconnected assemblies.

The carburettor is mounted within the right-hand crankcase cover, induction being controlled by a crankshaft-mounted disc valve assembly which times the induction cycle to achieve effective combustion.

## 2  Operations with engine/gearbox unit in frame

1  It is not necessary to remove the engine/gearbox unit from the frame in order to carry out the following service operations:

1  Removal and fitting of the cylinder head
2  Removal and fitting of the cylinder barrel
3  Removal and fitting of the piston assembly
4  Removal and fitting of the carburettor assembly
5  Removal and fitting of the rotary disc valve and valve seat assembly
6  Removal and fitting of the oil pump unit
7  Removal and fitting of the clutch assembly and primary drive pinion
8  Removal and fitting of the gearchange shaft and pawl mechanism
9  Removal and fitting of kickstart drive gear assembly
10  Removal and fitting of flywheel generator and contact breaker assembly
11  Removal and fitting of gearbox sprocket
12  Removal and fitting of neutral indicator switch

2  If it is found necessary to carry out several of these operations at the same time, it may be considered advantageous to remove the complete engine/gearbox unit from the frame in order to gain better access and more working space. This operation is comparatively simple and should take no more than an hour, whilst working at a leisurely pace without any assistance.

## 3  Operations with engine/gearbox unit removed from frame

1  Certain operations can be accomplished only if the complete engine unit is removed from the frame. This is because it is necessary to separate the crankcase to gain access to the parts concerned. These operations include:

1  Removal and fitting of the crankshaft assembly
2  Removal and fitting of the main bearings
3  Removal and fitting of the gearbox shaft and pinion assemblies, the gearbox bearings and the gear selector components
4  Removal and fitting of the kickstart shaft, spring and guide

## 4  Removing the engine/gearbox unit from the frame

1  Removal of the engine/gearbox unit will be made much easier by raising the machine to an acceptable working height and thus preventing the discomfort of squatting or kneeling down to work on the various component parts. Raising the machine may be achieved by using either a purpose built lift or a stout table or by building a platform from substantial planks and concrete blocks.

2  With the machine positioned on the chosen work surface and supported on its mainstand, place a container of at least 850 cc (1.50 Imp pint) capacity beneath the engine unit. Remove the drain plug from the underside of the right-hand crankcase half and allow the oil to drain from the gearbox. Note that this plug is in fact the housing for the detent plunger and spring and as the plug is removed, the plunger and spring will both be seen to follow it.

3  Whilst waiting for the gearbox oil to drain, examine and clean the detent plunger assembly. The spring must be renewed if it shows signs of fatigue or failure. The sealing washer should be renewed as a matter of course; always replace the used washer with a new one of an identical type, never omit to fit the washer or replace it with any form of sealant as this will cause the plunger assembly to malfunction because of the incorrect pressure placed upon the spring. Once cleaned, the assembly may be stored in a safe place ready for refitting. Note that if it is intended not to separate the crankcase halves then great care should be taken to avoid movement of the gearchange components whilst the detent plunger assembly is removed.

4  Remove the seat by releasing the two mounting bolts. These are located one either side towards the rear of the seat. The seat may then be lifted up and rearwards off its forward mounting point. It will be seen that these bolts also serve to support the rearmost section of rear mudguard. To avoid loss of the bolts, relocate them in the frame/seat attachment points whilst ensuring that each bolt has one serviceable spring

washer and the required number of plain washers located beneath its head.

5    Remove the fuel tank by first turning the fuel tap lever to the 'Off' position and releasing the fuel pipe retaining clip. This will allow the pipe to be pulled off the stub at the rear of the tap. Careful use of a small screwdriver may be necessary to help ease the pipe off the stub. Once the pipe is detached, allow any fuel in the pipe to drain into a small clean container. The tank may now be detached from the frame by unscrewing the single retaining bolt at the rear of the tank and pulling the tank up and rearwards off its front mounting rubbers. Place the tank retaining bolt, together with any associated mounting components in a safe place ready for refitting. Inspect the mounting rubbers for signs of damage or deterioration and if necessary, renew them before refitting of the tank is due to take place.

6    Detach each sidepanel from its frame mountings by unscrewing the single retaining screw located at the base of the panel and lifting the panel up to unclip it from its two upper attachment points. Carefully place the sidepanels, together with the seat and fuel tank, in a storage space where they are likely to be safe from damage. In the case of the fuel tank, the space must also be well ventilated and free from any source of naked flame or sparks.

7    Move to the right-hand side of the machine and isolate the battery from the electrical system by disconnecting the lead to the battery negative (-) terminal. This simple precaution will ensure that no shorting of exposed contacts occurs whilst disconnecting electrical components during removal of the engine unit. It is advisable at this stage to remove completely the battery from the machine so that it may be properly serviced in accordance with the instructions given in Chapter 6 of this Manual.

8    Trace the electrical leads from the flywheel generator stator assembly and disconnect them at their connection just to the rear of the seat mounting rubber, between the frame top tubes. Remove the clip retaining the leads to the frame downtube and thread the leads clear of the frame. Detach the suppressor cap from the spark plug and tuck both it and the HT lead clear of the cylinder head at a point on one of the frame top tubes. Loosen the spark plug now, as it is easier to free a really tight plug with the engine supported in the frame rather than with it loose on a work surface.

9    Prepare to drain the oil from the frame mounted oil tank by obtaining a clean container of at least 1.2 litre (0.4 Imp gal) and positioning it beneath the machine, as near to the base of the oil tank as possible. It will now be necessary to devise a method by which the oil draining from the stub at the base of the oil tank may be diverted into the container as soon as the oil feed pipe is removed. The obvious solution is to substitute for the oil feed pipe a similar length of pipe which may be routed from the tank stub to the container; otherwise a channel may be formed from a piece of clean card or similar material. Remove the oil feed pipe from the tank stub by releasing its retaining clip and carefully pulling it from position. Allow the contents of the tank to drain into the clean container.

10   Remove the gearchange lever by unscrewing its retaining bolt and pulling it clear of the gearchange shaft end. Detach the rear section of the left-hand crankcase cover by removing its four securing screws. With the gearbox sprocket exposed, the rear wheel should be rotated until the split link in the final drive chain appears between the chain guard and the gearbox sprocket whilst on its upper run. Removal of the chain may now be achieved by removing the spring clip of the split link with a pair of flat-nose pliers and withdrawing the link to allow the ends of the chain to separate. Clear the chain from the gearbox sprocket and allow its ends to rest on a piece of clean rag or paper placed beneath the machine. It may be considered beneficial at this stage, to remove completely the chain from the machine so that it may be properly cleaned, inspected for wear and relubricated.

11   Remove the two screws that hold the oil pump cover plate in position and manoeuvre the cover clear of the machine. Disconnect the pump control cable from the pump lever by

pushing the end of the lever up so that tension is taken off the cable inner and then detaching the cable nipple from its nylon holder. With the rubber sealing cap detached from the cable adjuster, the cable may now be pulled through the adjuster and clear of the machine.

12   Remove the contact breaker/clutch adjuster cover from the forward section of the left-hand crankcase cover by unscrewing its four retaining screws. Detach the crankcase cover by unscrewing its four retaining screws whilst noting, for reference when refitting, the position of the one short screw. Note that there is no need to detach the clutch operating cable from the lever within the cover unless the cover is to be replaced, the lever mechanism serviced or the cable renewed. Place the cover on a point of the frame clear of the engine unit. It is possible that the left-hand length of clutch pushrod will be withdrawn with the cover, in which case it should be detached from its location within the operating mechanism and stored safely ready for refitting. If the pushrod has remained in the gearbox mainshaft, it should now be removed before proceeding further.

13   The exhaust system should now be removed by first unscrewing the two bolts which serve to retain the exhaust pipe end to the cylinder barrel. Check that the spring washer located beneath each bolt head has not become flattened; if this is the case, the washers should be renewed before reassembly takes place. Remove the locknut from the end of the swinging arm fork pivot shaft and pull the silencer bracket clear of the shaft end. The complete exhaust system may now be manoeuvred clear of the machine by lifting the end of the silencer and threading the exhaust pipe end between the engine and footrest and rear brake pedal assemblies. Under no circumstances should the exhaust system be allowed to hang from the cylinder barrel mounting whilst disconnected from the swinging arm shaft as this will impose an unacceptable strain on the threads of the two mounting bolts.

14   Detach the cable seal retaining plate from the top of the carburettor housing by removing its four retaining screws. Work the seal, together with its retaining plate, up the throttle and choke cables until it is clear of the engine casing. If the seal shows signs of severe damage or deterioration, it must be renewed before reassembly of the engine takes place.

15   Remove the three screws from the carburettor cover and detach the cover. With the carburettor exposed, it will be seen that a total of five pipes are routed through the carburettor housing to connect at various points on the carburettor body. Although a photograph is provided with the text of this Chapter, it is important that the correct fitted position of each one of these pipes is noted for reference during reassembly; making a quick sketch is by far the best method of achieving this. Be prepared for any leakage of fuel when disconnecting the fuel feed pipe; have a small container or piece of rag handy in which to collect any fuel and observe the necessary fire precautions. Pull each pipe off its retaining stub, releasing retaining clips where necessary and making use of the flat of a small screwdriver to help ease any stubborn pipe from position.

16   Move to the front of the carburettor housing and remove the innermost of the two small blanking plugs. This will enable a screwdriver to be passed through the housing wall in order to slacken the carburettor mounting clamp. Pull the carburettor outwards off its mounting stub and manoeuvre it clear of the engine unit so that it can be placed on the frame top tubes. Check that neither the throttle nor choke cable will impede engine removal.

17   Disconnect the tachometer drive cable from the top of the gearbox housing by unscrewing its knurled retaining ring. Remove the air filter housing by loosening the inlet hose to crankcase retaining clamp and removing the two screws and washers holding the housing to its frame mounting. In practice, these two mounting screws were found to be very tight and required the use of an impact driver to free them. Manoeuvre the filter housing clear of the machine; if necessary, using the flat of a large screwdriver to ease carefully the hose end clear of the crankcase stub. Carry out a quick inspection of the inlet hose, it must be renewed if split or perished.

18 It should be noted at this point that Suzuki recommend removal of the cylinder head before proceeding further. This is a matter of personal preference, because in practice the head was left in position to protect the top threads of its securing stubs during engine removal, although the spark plug was removed to avoid any risk of its being knocked against the frame as the engine was lifted from position. Whichever course is decided upon, take note of the details given in Section 6 of this Chapter.

19 Remove each of the four engine unit mounting bolt nuts, noting whilst doing so the position of any washers for reference during reassembly. The engine is now ready for removal from the frame. The unit itself is light and can easily be lifted away by one person. It is helpful, however, to have an assistant present who can help to steady the machine and withdraw the mounting bolts. If the job is to be attempted single-handed arrange wooden blocks or a similar support beneath the engine crankcase to support the unit whilst the mounting bolts are withdrawn. Carry out a final check around the engine unit to ensure that no control cables, etc, remain connected or will impede the progress of the unit out of the frame. Withdraw the mounting bolts and lift the engine unit up slightly to clear it from the frame mounting plates before easing it away from the frame towards the right-hand side of the machine.

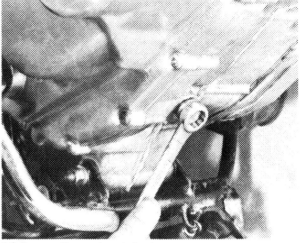
4.2 Remove the gearbox drain plug

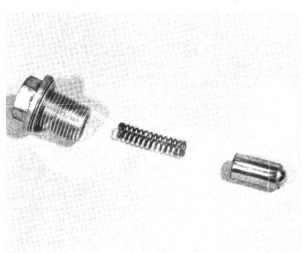
4.3 Examine the detent plunger assembly

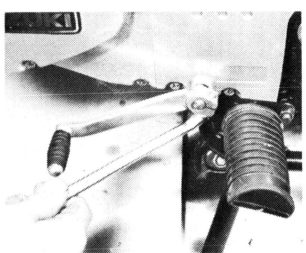
4.10a Remove the gearchange lever ...

4.10b ... to allow the removal of the gearbox sprocket cover

4.14 Detach the cable seal retaining plate ...

4.15a ... followed by the carburettor cover

4.15b Note the routing of the various pipes ...

4.16 ... before loosening the carburettor mounting clip

## 5 Dismantling the engine/gearbox unit: general

1 Before commencing work on the engine unit, the external surfaces should be cleaned thoroughly. A motorcycle engine has very little protection from road grit and other foreign matter, which will find its way into the dismantled engine if this simple precaution is not taken. One of the proprietary cleaning compounds, such as 'Gunk' or 'Jizer' can be used to good effect, particularly if the compound is worked into the film of oil and grease before it is washed away. Special care is necessary when washing down to prevent water from entering the now exposed parts of the engine unit.
2 Never use undue force to remove any stubborn part unless specific mention is made of this requirement. There is invariably good reason why a part is difficult to remove, often because the dismantling operation has been tackled in the wrong sequence.
3 Mention has already been made of the benefits of owning an impact driver. Most of these tools are equipped with a standard $\frac{1}{2}$ inch drive and an adaptor which can take a variety of screwdriver bits. It will be found that most engine casing screws will need jarring free due both to the effects of assembly by power tools and an inherent tendency for screws to become pinched in alloy castings. If an impact screwdriver is not available, it is often possible to use a crosshead screwdriver fitted with a T-handle as a substitute.
4 A cursory glance over many machines of only a few years' use, will almost invariably reveal an array of well-chewed screw heads. Not only is this unsightly, it can also make emergency repairs impossible. It should also be borne in mind that there are a number of types of crosshead screwdrivers which differ in the angle and design of the driving tangs. To this end, it is always advisable to ensure that the correct tool is available to suit a particular screw.
5 Before commencing dismantling, make arrangements for storing separately the various sub-assemblies and ancillary components, to prevent confusion on reassembly. Where possible, replace nuts and washers on the studs or bolts from which they were removed and refit nuts, bolts and washers to their components. This too will facilitate straightforward reassembly.

## 6 Dismantling the engine/gearbox unit: removing the cylinder head

1 Depending on the model type, the cylinder barrel will be retained by four special nuts or by four standard nuts, each with a plain washer. The cylinder head may be removed easily, irrespective of whether or not the engine unit is fitted in the frame.
2 To remove the cylinder head, slacken the four nuts evenly and in a diagonal sequence, this will avoid any risk of distortion. Remove the nuts and lift the head off its retaining studs. If the head appears to be stuck to the cylinder barrel, tap it lightly around its base with a soft-faced mallet; this should be sufficient to break any seal between the two components. Under no circumstances attempt to lever the head from position as this will only crack or distort the head casting.
3 With the head removed, discard the gasket. A new item should be fitted on reassembly. If washers are fitted, they should now be recovered from the head and stored along with the nuts.

## 7 Dismantling the engine/gearbox unit: removing the cylinder barrel, piston and small-end bearing

1 The cylinder barrel may be drawn off its retaining studs after the cylinder head has been removed. In practice, it was found that the barrel was stuck firmly to the crankcase and had

to be sharply tapped at the front and rear sections of strengthened finning with a length of wood and a mallet before it could be lifted clear. Under no circumstances should the areas of unsupported finning around the barrel be struck as this will only serve to break off the fins resulting in an area of barrel which will be inadequately cooled.

2    With the barrel freed from the crankcase, ease it gently upwards off its retaining studs. Take care to support the piston and rings as it emerges from the cylinder bore, otherwise there is risk of damage or ring breakage. If the crankcases are not to be separated, it is advisable to pack the crankcase mouth with clean rag before the piston is withdrawn from the bore, in case the piston rings have broken. This will prevent sections of broken ring from falling into the crankcase.

3    Prise one of the gudgeon pin circlips out of position, then press the gudgeon pin out of the small-end bearing through the piston boss. If the pin is a tight fit, it may be necessary to warm the piston so that the grip on the gudgeon pin is released. A rag

soaked in warm water and wrapped around the piston should suffice. The piston may be detached from the connecting rod once the gudgeon pin is clear of the small-end bearing.

4    If the gudgeon pin is still a tight fit after warming the piston, it can be lightly tapped out of position with a hammer and soft metal drift. **Do not** use excess force and make sure the connecting rod is supported during this operation, or there is a risk of its bending.

5    With the piston free of the connecting rod, remove the second circlip and fully withdraw the gudgeon pin from the piston. Place the piston and gudgeon pin aside for further attention. On no account reuse the circlips, they should be discarded and new ones fitted during rebuilding.

6    Push the small-end needle roller bearing out of the connecting rod eye and place it to one side, ready for cleaning and examination. Finally, remove and discard the cylinder barrel base gasket.

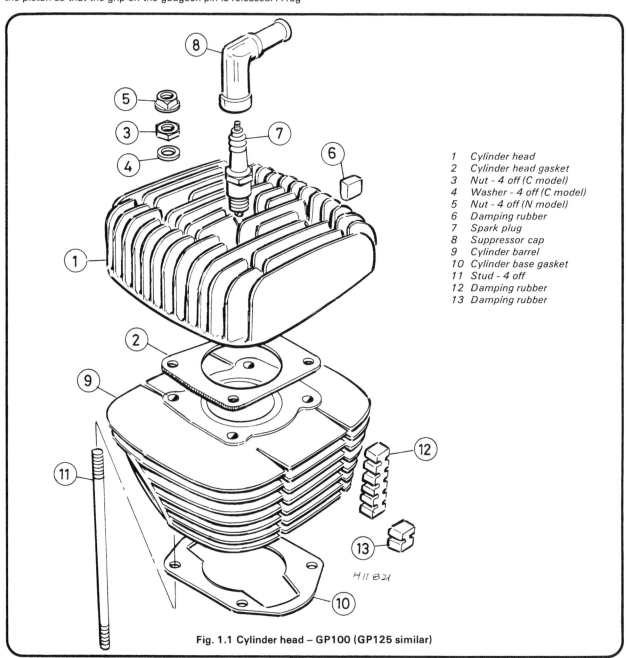

1   Cylinder head
2   Cylinder head gasket
3   Nut - 4 off (C model)
4   Washer - 4 off (C model)
5   Nut - 4 off (N model)
6   Damping rubber
7   Spark plug
8   Suppressor cap
9   Cylinder barrel
10  Cylinder base gasket
11  Stud - 4 off
12  Damping rubber
13  Damping rubber

Fig. 1.1 Cylinder head – GP100 (GP125 similar)

## 8 Dismantling the engine/gearbox unit: removing the flywheel generator assembly

1   The flywheel generator rotor is mounted on the left-hand end of the crankshaft; the stator assembly, which includes the contact breaker points, is attached to the crankcase beneath the rotor. The completed flywheel generator assembly may be removed with the engine in or out of the frame.

2   Before removal of the rotor can be attempted, the crankshaft must be prevented from rotating. This is achieved by passing a close-fitting bar through the small-end eye and allowing the ends of the bar to rest on wooden blocks placed on each side of the crankcase mouth. Never allow the ends of the bar to come into direct contact with the jointing face. If the rotor is to be removed with the engine in the frame, crankshaft rotation may be prevented by selecting top gear and applying the rear brake.

3   Remove the rotor retaining nut, spring washer and plain washer. The rotor is a tapered fit in the crankshaft end and is located by a Woodruff key; it therefore requires pulling from position. The rotor boss is threaded internally to take the special Suzuki service tool No 09930-30102 with an attachment, No 09930-30161. If this tool cannot be acquired, then it is possible to remove the rotor by careful use of a two-legged puller.

4   If it is decided to make use of a two-legged puller, great care must be taken to ensure that both the threaded end of the crankshaft and the rotor itself remain free from damage. To protect the crankshaft end, refit the rotor retaining nut and screw it on until its outer face lies flush with the end of the crankshaft. Assemble the puller, checking that its feet are fitted through the two larger slots in the rotor face and are resting securely on the strengthened hub. Remember that the closer the feet of the puller are to the centre of the rotor then the less chance there is of the rotor becoming distorted. Gradually tighten the centre bolt of the puller to apply pressure to the rotor. Do not overtighten this bolt but apply reasonable pressure and then strike the end of the bolt with a hammer to break the taper joint. If this fails at first, tighten the centre bolt to apply a little more pressure and then try shocking the rotor free again.

5   With the rotor removed from the crankshaft end, release the single electrical lead from its location on the neutral indicator switch. Using a sharp scribing tool or a centre punch, carefully mark the position of the stator plate in relation to the crankcase; this will serve as a reference when refitting the plate. Remove the three screws and washers that serve to retain the stator plate to the crankcase and lift the stator plate from position, threading the electrical leads clear of their retaining grommets whilst doing so.

6   Finally, ease the Woodruff key from its location in the crankshaft taper and place it in safe storage until required for reassembly.

## 9 Dismantling the engine/gearbox unit: removing the neutral indicator switch assembly and the gearbox sprocket

1   The neutral indicator switch takes the form of a white plastic cover which is sited over the left-hand end of the gearchange selector drum. To remove the switch, unscrew its two retaining screws and lift it clear of the selector drum end. It will be seen that an O-ring is fitted to the inside face of the switch, this should be discarded and replaced with a new item before refitting the switch. Using a pair of long-nose pliers, carefully withdraw the contact pin from its location in the end of the gearchange drum. Hook the spring out of the pin location and tilt the engine unit so that the ball upon which the spring seats rolls clear of the drum. Catch the ball and place it, together with the other switch components, in a clean, dry storage space. If the spring is seen to be damaged or seems to have become fatigued, then it must be replaced with a new

8.3 Remove the flywheel generator rotor by careful use of a two-legged puller

8.5 Mark the position of the generator stator plate

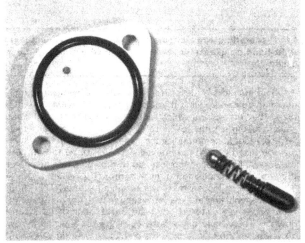

9.1 Carefully store the neutral indicator switch component parts

item. Note that it is possible that the ball bearing is not fitted as standard to all the model types covered in this Manual.

2   The gearbox sprocket is secured to the splined end of the gearbox layshaft by a large nut, the removal of which will necessitate locking the sprocket. With the engine in the frame, this is easily accomplished by applying the rear brake, thus immobilising the sprocket via the final drive chain. With the engine on the bench, it will be necessary to select fifth gear and lock the crankshaft in position by passing a close-fitting bar through the small-end eye and allowing the ends of the bar to rest on wooden blocks placed on each side of the crankcase mouth. Never allow the ends of the bar to come into direct contact with the jointing face.

3   Having locked the sprocket in position, knock back the tab of the lock washer from the nut and unscrew the nut. In practice, it was found that this was was in fact only finger-tight when it should have been tightened to quite a high torque loading. In theory, it may need some considerable effort to free the nut. Draw the sprocket off the layshaft, pull the spacer out the crankcase seal and note the position of the O-ring fitted behind the spacer.

## 10 Dismantling the engine/gearbox unit: removing the oil pump assembly

1   The oil pump is mounted in a compartment to the rear of the left-hand crankcase, and is retained by two mounting screws. The pump can be removed with the engine unit installed in, or removed from, the frame. In the case of the former, it will be necessary to detach the gearbox sprocket cover and oil pump cover, to gain access to the pump.

2   If the pump is to be removed with the engine in the frame, then provision must be made to catch the oil that will issue from both the feed and delivery pipes once they are disconnected from the pump. To prevent complete draining of the oil tank, the feed pipe should be plugged as soon as it is detached; a clean screw or bolt of the appropriate thread diameter is ideal for this purpose.

3   If the engine unit is removed from the frame, then the oil tank will have already been drained, in which case it is possible to leave both pipes connected to the oil pump during its removal. Slacken and remove both of the cross-headed retaining screws and lift the pump unit clear of its drive shaft end. Discard the base gasket. Replace it with a new item on reassembly.

## 11 Dismantling the engine/gearbox unit: removing the right-hand crankcase cover, primary drive pinion and clutch assembly

1   Free the kickstart lever from its shaft end by unscrewing its retaining bolt and pulling it free of its locating splines. It will be found that the screws which retain the cover to the crankcase are of varying lengths; because of this, it is well worth making up a cardboard template of the cover into which each screw may be inserted as it is removed. This will serve to both keep the screws from becoming lost and to provide a reference as to the correct fitted position of each screw.

2   Proceed to loosen the cover retaining screws. This should be done evenly and in a diagonal sequence to prevent the cover from becoming distorted. Pull the cover clear of its locating dowels, remove and discard the cover gasket and pull the two dowels from their location in the crankcase. The dowels should be placed in safe storage until required for reassembly.

3   Before attempting to dismantle the clutch assembly, it will first be necessary to manufacture a tool with which to ease the

pressure placed by the clutch springs on their anchor pins. Suzuki manufacture a special tool No 09920-20310 specifically for this purpose. It was found that a length of small gauge welding rod bent to the shape shown in the accompanying photograph was perfectly adequate for the job.

4   Proceed to release the pressure plate from the clutch assembly by placing the hooked end of the tool through the end of one of the clutch springs; pull the spring end out from the pressure plate so that it releases its grip on the anchor pin and draw the pin out through the spring end by using a pair of long-nose pliers. Repeat this procedure until all the pins are removed and the pressure plate can be lifted clear. Lift the clutch friction and plain plates out of the clutch drum as an assembly and place them on the work surface ready for examination. Remove the small central thrust bearing together with the pushrod end piece located beneath it. The main pushrod (GP100 C models), or two short pushrods (all other models), may now be withdrawn from the centre of the gearbox mainshaft.

5   It will now be necessary to devise some means of locking the gearbox mainshaft in position whilst the clutch hub retaining nut is slackened. The method used to do this was to lock the gearbox layshaft by utilising the gearbox sprocket, an old length of final drive chain, a length of metal tubing and a metal drift of a small enough diameter to pass through the links of the chain, and then to place the unit in gear. Commence by refitting the gearbox sprocket over the layshaft end and securing it in position by refitting its retaining nut, finger-tight. Engage the centre section of the length of chain over the sprocket and pass its ends through the length of tube. The metal drift should now be passed through both ends of the chain at a point where it will also abut against the end of the tube (see accompanying photograph).

6   With an assistant holding the layshaft locking tool in position, place the unit in gear and knock back the tab of the lock washer from the clutch hub retaining nut. Support the engine unit and slacken the retaining nut. In practice, this nut was found to be very tight and great care had to be taken to prevent the engine unit from rotating around its axis due to the opposing forces placed upon it. With the nut thus slackened, run it off the mainshaft thread and pull the clutch hub from position, together with the lock washer. Remove the thrust washer from the mainshaft, pull the clutch drum from position and remove the thick thrust washer located beneath it. Unscrew the clutch springs from their locations within the clutch hub and lay the complete clutch assembly out on the worksurface ready for examination.

7   The locking assembly should now be removed from the gearbox sprocket and the sprocket removed from the layshaft end. With this done, lock the crankshaft in position by passing a close-fitting bar through the small-end eye and allowing the ends of the bar to rest on wooden blocks placed on each side of the crankcase mouth. Never allow the ends of the bar to come into direct contact with the jointing face.

8   Knock back the tab of the lock washer from the primary drive pinion retaining nut and slacken the nut. Run the nut off the crankshaft thread, remove the lock washer, pull off the pinion and remove the Woodruff key.

## 12 Dismantling the engine/gearbox unit: removing the kickstart drive and idle pinions and the oil pump drive assembly

1   Remove the kickstart drive pinion from the kickstart shaft simply by lifting it up and off the shaft. Take note of the thrust washer which fits over the pinion for reference when refitting.

2   Remove the kickstart idle pinion by releasing its retaining circlip and drawing the pinion off its shaft. The oil pump drive pinion and shaft assembly may now be withdrawn from the crankcase.

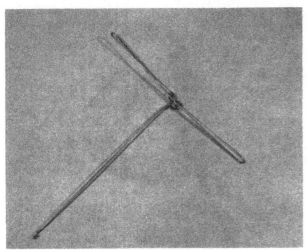

11.3 Make a clutch spring releasing tool ...

11.4 ... and with spring pressure eased, remove each anchor pin

11.5 Lock the gearbox layshaft in position

12.1a Note the thrust washer which fits over the kickstart drive pinion ...

12.1b ... before pulling the drive pinion off the kickstart shaft

**13 Dismantling the engine/gearbox unit: removing the gearchange shaft and pawl mechanism**

1    The gearchange shaft assembly can be removed from the crankcase by pulling it free of its spring locating pin and drawing it out of position.

2    The pawl mechanism can now be released from the crankcase by unscrewing the pawl lifter plate and cam guide pin securing screws. It was found in practice that all these screws were extremely tight and needed the use of an impact driver to free them. If it is found that only one of the two screws securing each plate can be freed with the driver, then moving the freed end of the plate may well serve to release the remaining screw. Take care when removing the pawl lifter plate to retain the two pawls in position in their housings, otherwise their operating springs will cause both the pawls and their operating pins to be ejected.

3    Remove the pawl mechanism and place it in safe storage until required for examination and reassembly. It is a good idea to wrap a rubber band around the unit to stop the pawls and pins from becoming separated and lost.

## 14 Dismantling the engine/gearbox unit: removing the disc valve assembly

1    Using an impact driver, loosen the five screws which retain the valve outer cover in position. With the screws removed, the cover can be lifted away together with the crankshaft oil seal.
2    It will be noted that a small timing mark is made in the edge of the valve plate boss. This mark takes the form of a small notch and should be in direct alignment with the crankshaft keyway. To preclude the risk of mistiming the valve during reassembly, carefully note the position of the timing mark and also degrease a small area of the outward facing surface of the valve so that it may be marked with an indelible pen. Carrying out these simple tasks will save much frustration at a later stage and obviate completely the risk of the valve being fitted upside down. Remove the valve plate.
3    Again, using an impact driver, loosen and remove the five screws which retain the valve seat to the crankcase. It may be found that the seat has become stuck in position and cannot be easily lifted away, in which case the flat of a small screwdriver should be inserted into the slot provided between the outer face of the crankcase inlet bore and the inner face of the valve seat and gentle pressure applied to the screwdriver until the two components part.
4    With the valve seat removed, the central splined boss over which the valve plate is fitted can now be slid off the crankshaft end. It is possible that this component will also prove difficult to remove due to the fact that sealing compound will have been applied to its mating surface with the crankshaft bearing. If this is so, the boss will have to be freed by carefully using the flat of a screwdriver to lever it from position whilst taking great care to avoid damaging the surrounding aluminium alloy components. Remove and discard the valve seat base gasket.

14.3 Note the leverage point (arrowed) provided in the crankcase inlet bore

## 15 Dismantling the engine/gearbox unit: separating the crankcase halves

1    Prepare the crankcase for separation by arranging it on the work surface so that it is well supported on wooden blocks with its left-hand side facing uppermost. It will be found that the twelve screws which retain the crankcase halves together are of various lengths. To prevent these screws from becoming refitted in the wrong locations, it is a good idea to make up a cardboard template of the crankcase through which the screws can be fitted as they are removed. An impact driver will be required to free the screws before they can be removed.
2    When removing these screws, take care to work in a diagonal sequence, loosening the screws in small increments at first until they are all completely free. This simple precaution will obviate any risk of the crankcase halves becoming distorted.
3    Check that all twelve screws are removed and attempt to lift the left-hand crankcase half off the right-hand assembly. The crankcase halves have been sealed along their mating surfaces with a sealing compound and it is likely that this compound will prevent easy separation of the two components. It was found that tapping around the joint with a soft-faced mallet served to break this seal and once this was done, the crankcase halves were easily pulled apart.
4    Resist the temptation to lever the casing halves apart with a screwdriver, as this almost invariably leads to damaged sealing faces. It is particularly important that the crankcase joint remains absolutely airtight on a two-stroke engine, as an air leak can cause loss of secondary compression, and can upset the fuel-air mixture. Remove the two large locating dowels from their mating surface locations and place them in safe storage until required for reassembly.

## 16 Dismantling the engine/gearbox unit: removing the kickstart shaft assembly, gearbox components, tachometer drive gear and crankshaft

1    It will be found that once the crankcase halves are separated, the crankshaft assembly will almost always remain with the left-hand crankcase half whilst the gearbox components and kickstart shaft assembly remain within the right-hand crankcase half. With the right-hand crankcase half supported on the work surface as for crankcase separation, remove the plastic guide from within the kickstart return spring and unlatch the return spring from its location in the shaft by using a pair of long-nose pliers. Remove the spring, rotate the shaft so that its ratchet stop clears its retaining plate and draw the shaft from position.
2    The detent plunger assembly will have already been removed from the right-hand crankcase half for the purpose of draining the gearbox oil. Remove the gearchange drum by freeing it from the selector fork pins and pulling it out of its location. It is well worth making a quick sketch of the fitted position of all three selector forks before they are disturbed; this simple precaution will save considerable anxiety at a later stage if the forks should become separated from their shafts during the cleaning and examination procedures.
3    Withdraw each selector fork shaft, detaching the fork(s) from the gear clusters and relocating each one in its correct fitted position on the shaft directly after doing so. Place the reassembled components on a clean worksurface.
4    Withdraw both the mainshaft and layshaft assemblies from the crankcase half as a complete unit, taking great care to keep both assemblies together. Place the complete unit next to the selector fork shafts and cover all the gearbox components with a piece of clean rag or paper to prevent the ingress of dirt into the bearing surfaces. If none of the assemblies are to be dismantled, it is advisable to secure them in their correct relative positions with elastic bands before proceeding further. The right-hand crankcase half is now free of all removable components except the mainshaft bearing retainer plate, the kickstart ratchet stop retaining plate, the gearchange shaft return spring stop and the tachometer drive gear. Of these three components, only the tachometer drive gear needs to be removed, for the purposes of inspection and lubrication.
5    To remove the tachometer drive gear, unscrew the single

retaining screw from the side of the drive gear housing and withdraw the complete assembly from position. Relocate the retaining screw in the housing and tighten it finger-tight so that it does not become separated from the crankcase until required for reassembly. Note that there should be a plain washer fitted beneath the head of this screw.

6    Both the mainshaft bearing retainer plate and the kickstart ratchet stop retaining plate are held to the crankcase by two retaining screws. In the case of the latter component, the screw heads are locked in position by a tab washer; this washer must be renewed once the screws are removed.

7    The right-hand crankcase half should now be placed to one side and the left-hand half supported on wooden blocks on the work surface so that the crankshaft can be pushed clear of its bearing without its end coming into contact with the work surface. Great care must be taken to ensure that the crankcase half is well supported at all points on its mating surface, as the crankshaft will have to be struck quite sharply to free it and this simple precaution will lessen the risk of the casing being distorted.

8    The method used to free the crankshaft was to place a heavy copper drift against the crankshaft and sharply strike the drift with a heavy hammer. It was found that the crankshaft moved on the first blow and continued to move under the force of subsequent lighter blows until completely free. In order to protect the end of the crankshaft during this procedure, the flywheel generator rotor retaining nut was refitted and run down its thread until the end of the shaft appeared flush with the upper side of the nut.

9    If it is thought that excessive force is being required to move the crankshaft and there is a serious risk of the crankcase becoming damaged, it is recommended that advice and aid be sought from an official Suzuki service agent. Alternatively, Suzuki provide a service tool No 09920-13111 for pressing the shaft from position.

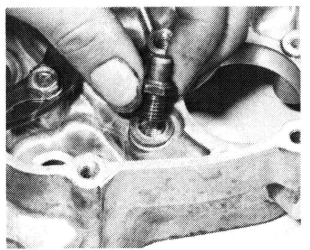

16.4a The gearchange shaft return spring stop ...

## 17  Examination and renovation: general

1    Before examining the parts of the dismantled engine unit for wear, it is essential that they should be cleaned thoroughly. Use a paraffin/petrol mix to remove all traces of old oil and sludge that may have accumulated within the engine.

2    Examine the crankcase castings for cracks or other signs of damage. If a crack is discovered, it will require professional repair.

3    Examine carefully each part to determine the extent of wear, if necessary checking with the tolerance figures listed in the Specifications Section of this Chapter, or accompanying the text.

4    Use a clean, lint-free rag for cleaning and drying the various components, otherwise there is risk of small particles obstructing the internal oilways.

5    Should any studs or internal threads require repair, now is the appropriate time to attend to them. Where internal threads are stripped or badly worn, it is preferable to use a thread insert, rather than tap oversize. Most dealers can provide a thread reclaiming service by the use of Helicoil thread inserts. They enable the original component to be re-used.

6    Sheared studs or screws can usually be removed with screw extractors, which consist of tapered, left-hand thread screws, of very hard steel. These are inserted by screwing anticlockwise, into a pre-drilled hole in the stud, and usually succeed in dislodging the most stubborn stud or screw. The only alternative to this is spark erosion, but as this is a very limited, specialised facility, it will probably be unavailable to most owners. It is wise, however, to consult a professional engineering firm before condemning an otherwise sound casing. Many of these firms advertise regularly in the motorcycle papers.

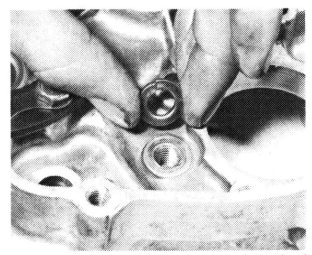

16.4b ... must have a serviceable spring washer (where fitted)

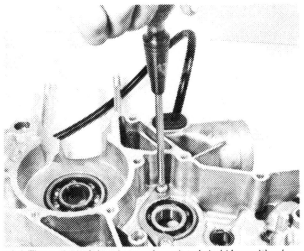

16.6 The mainshaft bearing retainer plate is held in position by two screws

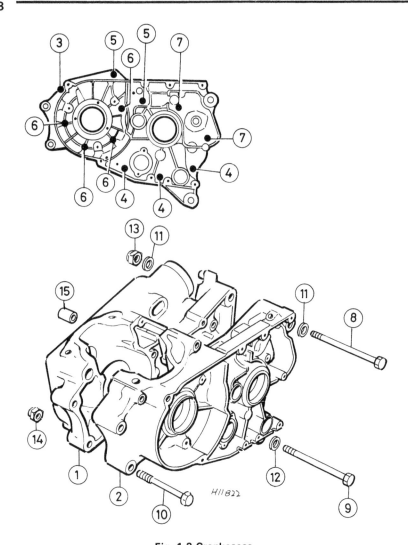

1   Right-hand crankcase half
2   Left-hand crankcase half
3   Screw
4   Screw - 3 off
5   Screw - 5 off
6   Screw
7   Screw - 2 off
8   Bolt
9   Bolt
10  Bolt - 2 off
11  Washer
12  Washer - 2 off
13  Nut - 2 off
14  Nut - 2 off
15  Hollow dowel - 2 off

H11822

**Fig. 1.2 Crankcases**

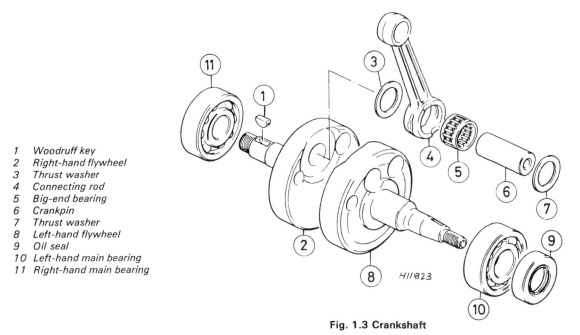

1   Woodruff key
2   Right-hand flywheel
3   Thrust washer
4   Connecting rod
5   Big-end bearing
6   Crankpin
7   Thrust washer
8   Left-hand flywheel
9   Oil seal
10  Left-hand main bearing
11  Right-hand main bearing

H11823

**Fig. 1.3 Crankshaft**

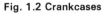

## 18 Examination and renovation: crankcase and gearbox oil seals

1   Any failure in the crankcase oil seals on a two-stroke engine will have a profound effect on the performance and efficiency of the unit. When the seals begin to wear, air is admitted into the crankcase; this air serves to dilute the incoming fuel/air mixture, thereby causing uneven running and difficulty in starting.

2   Examine each seal carefully, paying particular attention to its sealing lip. This area performs the sealing function, and the seal should be renewed without question if it is scored or marked in any way. If the seal is found to have become hardened, due to the machine having been in use for a considerable amount of time, it should again be renewed without question. In view of the important part that every seal plays, it is considered good practice to renew them as a set as a matter of course whilst the engine is dismantled.

3   Each seal can be removed by prising it out of position with the flat of a screwdriver. Great care should be taken, however, to ensure that the alloy in or around the seal housings is not damaged during this operation. Equal care should be taken when fitting a new seal. Ensure that the seal is facing in the right direction and push it as far as possible into its housing with hand pressure whilst ensuring that it enters squarely. Select a socket or length of metal tube that has an external diameter slightly less than that of the seal and with the casing well supported around the seal housing, gently drift the seal into position. Note that the two crankcase oil seals must have a thread locking compound applied to their outer edges before insertion otherwise they will tend to rotate out of position during use.

## 19 Examination and renovation: crankshaft main bearings and gearbox bearings

1   Failure of the crankshaft main bearings is usually evident in the form of an audible rumble from the bottom end of the engine, accompanied by vibration. The vibration will be most noticeable through the footrests. Main bearing failure will immediately be obvious when the bearings are inspected, after the old oil has been washed out. If any play is evident or if the bearings do not run freely, renewal is essential.

2   Each bearing can be removed by first removing its oil seal or retaining plate, where appropriate, and then applying heat to the engine casing. The application of heat will cause the aluminium alloy of the casing to expand at a faster rate than the steel of the bearing thus allowing the bearing to become loose. The safest way of doing this is to place the casing in an oven, heating it to approximately 80-100°C, or to immerse the casing in boiling water. The casing may then be sharply tapped on a wooden surface, face down, to jar the bearing free. Alternatively, the bearing may be drifted out of position by using a hammer in conjunction with a length of metal tube of the appropriate diameter; take care when doing this to ensure that the casing is properly supported and that the bearing is made to leave the casing squarely. Note that care should always be exercised when heating alloy casings as excessive or localised heat can easily cause warpage.

3   Each new bearing can be drifted into position, after heating the casing as for removal, by using a socket or length of metal tube. The diameter of the tube should be equal to that of the outer race of the bearings. It is essential that the component to which the bearing is to be fitted is well supported during the fitting procedure and that the bearing enters the casing squarely. Check, before fitting the left-hand bearing, that the oil feed drilling is clear of any sludge or contamination.

4   The procedure for examination, removal and fitting of the gearbox bearings is as listed in the above paragraphs of this Section. Note that the gearbox layshaft bush, which is fitted in the right-hand crankcase half, may be dealt with in a manner similar to that described for the ball bearings.

18.3a Prise each defective seal from position

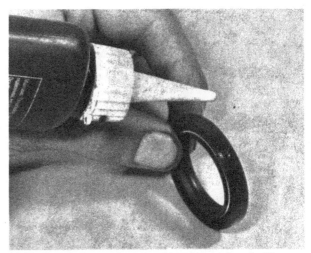

18.3b Coat the outer edge of each crankcase oil seal with thread locking compound ...

18.3c ... before fitting the seal into its location

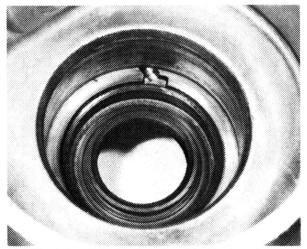

19.3 Check that the bearing housing oil feed is clear of contamination

19.4a Drive the gearbox layshaft bush into position ...

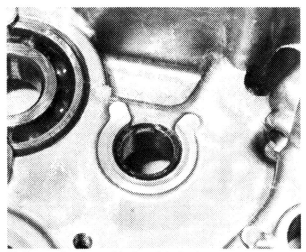

19.4b ... noting the positioning of the bush cutout

## 20 Examination and renovation: crankshaft assembly

1    The crankshaft assembly comprises two full flywheels, two mainshafts, a crankpin and big-end bearing, a connecting rod, and a caged needle roller small-end bearing. The general condition of the big-end bearing may be established with the assembly removed from the engine, or with just the cylinder head and barrel removed, as would be the case during a normal decoke. In this way it is possible to decide whether big-end renewal is necessary, without a great deal of exploratory dismantling.

2    Big-end failure is characterised by a pronounced knock which will be most noticeable when the engine is worked hard. The usual causes of failure are normal wear, or a failure of the lubrication supply. In the case of the latter, big-end wear will become apparent very suddenly, and will rapidly worsen. Check for wear with the crankshaft set in the TDC (top dead centre) position, by pushing and pulling the connecting rod. No discernible movement will be evident in an unworn bearing, but care must be taken not to confuse end float, which is normal, and bearing wear. If a dial gauge is readily available, a further test may be carried out by setting the gauge pointer so that it abuts against the upper edge of the periphery of the small-end eye. Measurement may then be taken of the amount of side-to-side deflection of the connecting rod. If this measurement exceeds the service limit of 3.0 mm (0.12 in) then the big-end bearing must be renewed.

3    Any measurement of crankshaft deflection can only be made with the crankshaft assembly removed from the crankcase and set up on V-blocks which themselves have been positioned on a completely flat surface. The amount of deflection should be measured with a dial gauge at a point just inboard of the threaded portion of the left-hand mainshaft. If the amount of deflection shown by the gauge needle exceeds the service limit of 0.05 mm (0.002 in), then the assembly must be replaced with a serviceable item.

4    Like the big-end bearing, the small-end bearing is of the caged needle roller type and will seldom give trouble unless a lubrication failure has occurred. Push the bearing into the small-end eye of the connecting rod and push the gudgeon pin through the bearing. Hold the connecting rod steady and feel for any discernible movement between it and the gudgeon pin. If movement is felt, do not automatically assume that the bearing is worn but check that the bore of the small-end eye and the outer diameter of the gudgeon pin are not worn beyond their service limits. If both the connecting rod and gudgeon pin are found to be within limits, discard the bearing and replace it with a new item. Close inspection of the bearing will show if the roller cage is beginning to crack or wear, in which case the bearing must be renewed.

5    If any fault is found or suspected in any of the components comprising the crankshaft assembly, it is recommended that the complete crankshaft assembly is taken to a Suzuki Service Agent, who will be able to confirm the worst, and supply a new or service-exchange assembly. The task of dismantling and reconditioning the big-end assembly is a specialist task, and it is considered to be beyond the scope and facilities of the average owner.

## 21 Examination and renovation: decarbonising

1    Decarbonising must take place as part of any major overhaul, in addition to being a normal routine maintenance function. In the case of the latter, the operation can be undertaken with minimal dismantling, namely removal of the cylinder head, cylinder barrel and exhaust silencer baffle. Carbon build up in a two-stroke engine is more rapid than that of its four-stroke counterpart, due to the oily nature of the combustion mixture. It is however, rather softer and is therefore more easily removed.

2 The object of the exercise is to remove all traces of carbon whilst avoiding the removal of the metal surface on which it is deposited. It follows that care must be taken when sealing with the relatively soft alloy cylinder head and piston. Never use a steel scraper or screwdriver for carbon removal. A hardwood, brass or aluminium scraper is the ideal tool as these are harder than the carbon, but no harder than the underlying metal. Once the bulk of the carbon has been removed, a brass wire brush of the type used to clean suede shoes can be used to good effect.

3 The whole of the combustion chamber should be cleaned, as should the piston crown. It is recommended that as smooth a finish as possible is obtained, as this will show the subsequent build up of carbon. If desired metal polish can be used to obtain a smooth surface. The exhaust port must also be cleaned out, as a build up of carbon in this area will restrict the flow of exhaust gases from the cylinder. Take care to remove all traces of debris from the cylinder and ports, prior to reassembly, by washing the components thoroughly in petrol or paraffin whilst taking care to observe the necessary fire precautions.

4 Full details of decarbonising the silencer assembly are contained in Section 16 of the following Chapter.

## 22 Examination and renovation: cylinder head

1 Check that the cylinder head fins are not clogged with oil or road dirt, otherwise the engine will overheat. If necessary, use a degreasing agent and brush to clean between the fins. Check that no cracks are evident, especially in the vicinity of the holes through which the holding down studs and bolts pass, and near the spark plug threads.

2 Check the condition of the thread in the spark plug hole. If it is damaged an effective repair can be made using a Helicoil thread insert. This service is available from most Suzuki Service Agents. The cause of a damaged thread can usually be traced to overtightening of the plug or using a plug of too long a reach. Always use the correct plug and do not overtighten.

3 If leakage problems have been experienced between the cylinder head and cylinder barrel mating surfaces, the cylinder head should be checked for distortion by placing a straight-edge across several places on the mating surface and attempting to slide a 0.05 mm (0.002 in) feeler gauge between the straight-edge and the mating surface.

4 If the cylinder head is found to be warped beyond this limit, grind it flat by placing a sheet of emery paper on a surface plate or sheet of plate glass and rubbing the cylinder head mating surface against it, in a slow, circular motion. Commence with 200 grade paper and finish with 400 grade paper and oil. If it is found necessary to remove a substantial amount of metal before the mating surface becomes completely flat, obtain advice from a Suzuki service agent as to whether it is necessary to obtain a new head.

20.1 The crankshaft assembly

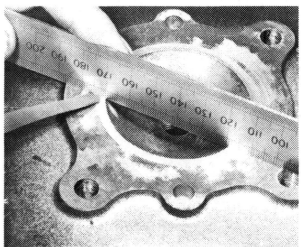

22.3 Check the cylinder head for distortion

5 Note that most cases of cylinder head distortion can be traced to unequal tensioning of the cylinder head securing nuts and by tightening them in the incorrect sequence.

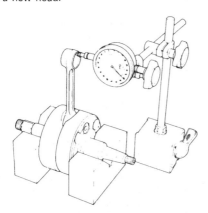

Fig. 1.4 Measuring the small-end deflection

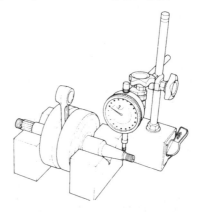

Fig. 1.5 Measuring the amount of crankshaft deflection

## 23 Examination and renovation: cylinder barrel

1    The usual indication of a badly worn cylinder barrel and piston is known as piston slap, a metallic rattle that occurs when there is little or no load on the engine.

2    Commence by cleaning the outside of the cylinder barrel, taking care to remove any accumulation of dirt from between the cooling fins. Carefully remove the ring of carbon from the mouth of the bore, so that an accurate assessment of bore wear can be made.

3    A close visual examination of the bore surface must be made, to check for scoring or any other damage, particularly if broken piston rings were encountered during the stripdown. Any damage of this nature will necessitate reboring and a new piston, as it is impossible to obtain a satisfactory seal if the bore is not perfectly finished.

4    There will probably be a lip at the uppermost end of the cylinder bore which marks the limit of travel of the top of the piston ring. The depth of the lip will give some indication of the amount of bore wear that has taken place even though the amount of wear is not evenly distributed.

5    The most accurate method of measuring bore wear is by the use of a cylinder bore DTI (Dial Test Indicator) or a bore micrometer. Suzuki supply a special tool, No 09900-20508, specifically for this purpose. Measurement should be taken at a point 15 mm (0.59 in) from the top of the cylinder bore and the reading obtained compared with the service limit for bore wear given in the Specifications Section of this Chapter. Note that it is necessary to rotate the measuring instrument so that the point of greatest wear in the bore is found.

6    It is however, most unlikely that the average owner will have the above listed items of equipment readily available. A slightly less accurate but more practical method is as follows. It is possible to determine the amount of bore wear by inserting the piston, without rings, so that it is just below the ridge at the top of the bore. Measure the distance between the bore wall and the side of the piston using feeler gauges. Move the piston down to the bottom of the bore and repeat the measurement. Doing this, and subtracting the lesser measurement from the greater, will give the difference between the bore diameter in an area where the greatest amount of wear is likely to occur and an area in which there should be little or no wear. If the difference found exceeds 0.10 mm (0.004 in), then remedial action will have to be taken. Note that the curvature of the gap being measured will tend to preclude accurate measurement by this method but it should be possible to gain a good indication of whether a rebore is necessary.

7    If it is found that the amount of wear in the bore exceeds those limits given, then it will be necessary to have the cylinder barrel rebored to the next oversize and the appropriate oversize piston fitted. Suzuki supply pistons in two oversizes: +0.5 mm (0.196 in) and +1.0 mm (0.0394 in).

8    On receiving the cylinder barrel back from the service agent who has carried out the rebore, check the edges of the ports at the bore end to ensure that they have been correctly chamfered to the measurements given in the figure accompanying this text. This work must be done before the cylinder barrel is refitted to the machine, otherwise there is a distinct possiblity that the piston rings will catch on the unchamfered edge of each port and break, thus necessitating a further strip down and rebore. Chamfering of the port edges can be carried out by very careful use of a scraper but it is essential to ensure that the wall of the bore does not become damaged in the process. Finish off the process by polishing the cut edges with fine emery paper.

9    Establishment of the clearance between the cylinder bore and the piston can be made either by direct measurement of the cylinder bore and piston diameters and then by subtracting the latter figure from the former, or by actual measurement of the gap using a feeler gauge. In either case, if the clearance exceeds the maximum wear limit there is evidence that a new piston or a rebore and new piston is required. Note that if the method of direct measurement of the piston and bore is decided upon,

then the measurement for piston diameter should be made at a point 20 mm (0.79 in) from the base of the piston skirt, at right-angles to the gudgeon pin hole, whereas the measurement for cylinder bore diameter should be made at various points around a line 20 mm (0.79 in) from the top of the bore.

10  Finally, check the cylinder barrel to cylinder head mating surface for distortion by placing a straight-edge across several places on it and attempting to slide a 0.05 mm (0.002 in) feeler gauge between the straight-edge and mating surface. If the cylinder barrel proves to be warped beyond this limit, grind it flat by placing a sheet of emery paper on a surface plate or sheet of plate glass and rubbing the mating surface against it, in a slow, circular motion. Commence this operation with 200 grade paper and finish with 400 grade paper and oil. If it is thought necessary to remove a substantial amount of metal in order to bring the mating surface back to within limits, obtain advice from a Suzuki service agent as to whether it is necessary to obtain a replacement barrel.

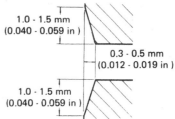

1.0 - 1.5 mm
(0.040 - 0.059 in )

0.3 - 0.5 mm
(0.012 - 0.019 in )

1.0 - 1.5 mm
(0.040 - 0.059 in )

**Fig. 1.6 Cylinder bore port chamfer measurements**

## 24 Examination and renovation: piston and piston rings

1    If a rebore is necessary, the existing piston and rings can be disregarded because they will be replaced with their oversize equivalents as a matter of course.

2    Remove the piston rings by pushing the ends apart with the thumbs whilst gently easing each ring from its groove. Great care is necessary throughout this operation because the rings are brittle and will break easily if overstressed. If the rings are gummed in their grooves, three strips of tin can be used, to ease them free, as shown in the accompanying illustration. Take great care to keep the rings separate and the right way up so that they can be refitted in their correct positions.

3    Piston wear usually occurs at the skirt or lower end of the piston and takes the form of vertical streaks or score marks on the thrust side. There may also be some variation in the thickness of the skirt. Measurement for piston diameter should be taken at a point 20 mm (0.79 in) from the base of the piston skirt, at right-angles to the gudgeon pin hole. If the measurement obtained is found to be less than the service limit given in the Specifications Section of this Chapter, then the piston must be renewed.

4    Check that the piston and bore are not scored, particularly if the engine has tightened up or seized. If the bore is badly scored, it will require a rebore and oversize piston. If the scoring is not too severe or the piston has just picked up, it is possible to remove the piston high spots by careful use of a fine swiss file. Application of chalk to the file teeth will help prevent clogging of the teeth, and the subsequent risk of scoring.

5    Check for any build up of carbon in the piston ring grooves. Any carbon should be carefully removed by using a section of broken piston ring or similar. The piston ring grooves may have become enlarged in use, thus allowing each ring to have a greater side float than is permissible. To measure this side float, insert each piston ring in its cleaned groove and measure the clearance between the side of the ring and the groove with a feeler gauge. If the measurement obtained exceeds 0.03 – 0.07 mm (0.001 – 0.003 in), then the piston is due for renewal.

6    When cleaning the piston ring grooves, check that the ring locating peg located in each groove is not loose or worn. If in doubt as to the condition of these pegs, seek professional

advice from a Suzuki service agent and renew the piston if necessary.

7  Examine the gudgeon pin for any scoring on its bearing surface. If damage is apparent on its contact surface with the holes in the piston, examine the surface of these holes and reject the piston if similar damage is found. The gudgeon pin should be a firm press fit in the piston. If any excessive looseness is felt between the two components, check the pin to piston clearance and compare the reading obtained with the service limit given in the Specifications Section of this Chapter. If this reading is found to be excessive, measure the gudgeon pin outer diameter and the piston hole bore diameter before rejecting one or both of the components, as necessary.

8  Check that the gudgeon pin circlip retaining grooves in the piston are free from damage. If there is the slightest doubt as to the condition of these grooves, seek further advice from a Suzuki service agent who will determine whether or not the piston should be renewed. Remember that if one of these rings should become detached in use, certain damage will be caused to both the piston and the surface of the bore.

9  Examine the working surface of each piston ring. If discoloured areas are evident, the ring should be renewed because these areas indicate the blow-by of gas.

10  Piston ring wear is measured by inserting each ring in a part of the cylinder bore which is not normally subject to wear. Ideally, the ring should be inserted into the cylinder bore so that it is positioned approximately 10 mm (0.4 in) from the bore spigot cutouts. Use the crown of the piston as a means of locating the ring squarely in the bore and measure the gap between the ring ends with a feeler gauge. If the gap measured is found to be greater than the service limit of 0.80 mm (0.031 in), then the ring must be renewed. Note that ring end gap service limits are given for when each ring is in its free state. These are given as a means of determining the spring tension of each ring. Measure this end gap with the ring placed flat on a clean work surface and compare the reading obtained with the appropriate reading given in the Specifications Section of this Chapter. The piston rings are marked on their upper surface by the letter 'T' or 'N' depending on the manufacturer.

11  It must be appreciated that, if a set of new rings is to be fitted to a used piston which is to run in a part worn bore, then the upper of the two rings should be stepped to clear the ridge left in the bore by the previous top ring. If a new ring is fitted, it will hit the ridge and break. This is because the new ring will not have worn in the same way as the old, which will have worn in unison wth the ridge.

12  Suzuki do not supply a stepped upper ring to fit any of the machines covered in this Manual. It is therefore necessary to ensure that the wear ridge at the top of the bore is completely removed before fitting a set of new rings. This may be incorporated in the 'glaze busting' process, which should be carried out in order to break down the surface glaze on the used bore so that the new rings are able to bed in to the bore surface. This combined process of ridge removal and 'glaze busting' can be carried out by any one of many competent engineering firms who advertise regularly in the national motorcycle press; your local Suzuki service agent could well offer a similar service.

13  Bear in mind, when refitting used rings to the piston, that they must be fitted in exactly the same positions as noted during removal. When fitting either new or old rings, take note of the following points. Each ring should be fitted to the piston by pulling its ends apart just enough to allow it to pass over the piston crown and into its groove. Always fit each top and second ring with its marked (T or N) surface uppermost; the two rings are identical when new. Check that each ring presses easily into its groove and that its ends locate correctly over the locating pin.

14  Finally, do not automatically assume when fitting new rings, that their end gaps will be correct. As with part worn rings, the end gap must be measured. It may be necessary to enlarge the gap, in which case this should be done by careful use of a needle file. With the rings fitted to the piston, place the assembly to one side ready for engine reassembly.

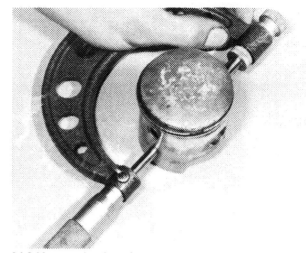

24.3 Measure the piston for wear

24.5 Measure the piston ring to groove clearance

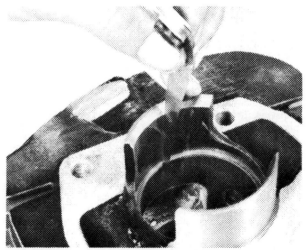

24.10a Position each ring in the cylinder bore to measure ring wear

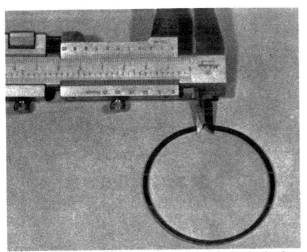

24.10b Check the spring tension of each ring by measuring its end gap

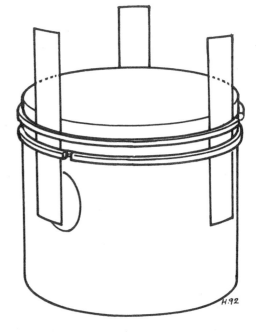

Fig. 1.7 Method of removing and replacing piston rings

---

## 25 Examination and renovation: gearbox components

1   Examine each of the gear pinions to ensure that there are no chipped or broken teeth and that the dogs on the end of the pinions are not rounded. Gear pinions with these defects must be renewed; there is no satisfactory method of reclaiming them. If damage or wear warrants renewal of any gear pinions the assemblies may be stripped down, displacing the various shims and circlips as necessary.

2   If dismantling of the mainshaft assembly proves to be necessary, the 2nd gear pinion will have to be pressed from position by using a hydraulic press; no other method of removal is possible. As it is unlikely that this type of tool will be readily available, it is recommended that the complete mainshaft assembly be returned to an official Suzuki service agent who will be able to remove the pinion, renew any worn or damaged components and return the mainshaft assembly complete.

3   If a hydraulic press is available and it is decided to attempt removal of the 2nd gear pinion from the mainshaft, it is very important to fully realise the dangers involved when using such a tool. Both the tool and the mainshaft assembly must be set up so that there is no danger of either item slipping. The tool must be correctly assembled in accordance with the maker's instructions as the force exerted by the tool is considerable and perfectly capable of stripping any threads from holding studs or inflicting other damage upon itself and the mainshaft. Always wear proper eye protection in case a component should fail and shatter before it becomes free, as will happen if the component is flawed. The type of press used in practice is shown in the photograph accompanying this text; some considerable force was needed to break the seal between the pinion and the shaft due to the application of a locking compound between the two components during assembly. To aid removal, purchase with the tool should be made behind the 4th gear pinion which is the large diameter pinion adjacent to the 2nd gear pinion. Both pinions will be drawn off at the same time.

4   The accompanying illustrations show how both clusters of the gearbox are assembled on their respective shafts. It is imperative that the gear clusters, including the thrust washers, are assembled in **exactly** the correct sequence, otherwise constant gear selection problems will occur. In order to eliminate the risk of misplacement, make rough sketches as the clusters are dismantled. Also strip and rebuild as soon as possible to reduce any confusion which might occur at a later date.

5   Examine the selector forks carefully, ensuring that there is no sign of scoring on the bearing surface of either their fork ends, their bores or their gearchange drum locating pins. Check for any signs of cracking around the edges of the bores or at the base of the fork arms. Refer to the Specifications Section at the beginning of this Chapter and measure the thickness of the fork claw ends; renew the fork if the measurement obtained is less than the limit given.

6   Place each selector fork in its respective pinion groove and, using a feeler gauge, measure the claw end to pinion groove clearance. If the measurement obtained exceeds the given service limit of 0.45 mm (0.018 in), then it must be decided whether it is necessary to renew one or both components. The acceptable limits for pinion groove width are given in the Specifications Section of this Chapter.

7   Check each selector fork shaft for straightness by rolling it on a sheet of plate glass and checking for any clearance between the shaft and the glass with feeler gauges. A bent shaft will cause difficulty in selecting gears. There should be no sign of any scoring on the bearing surface of the shaft or any discernible play between each shaft and its selector fork(s).

8   The tracks in the gearchange drum, which co-ordinate the movement of the selector forks, should not show signs of undue wear or damage. Check also that the contact surfaces of the detent plunger and drum are not worn or damaged.

9   Note that certain pinions have a bush fitted within their centres. If any one of these bushes appears to be overworn or in any way damaged, then the pinion should be returned to an official Suzuki service agent, who will be able to advise on which course of action to take as to its renewal.

10  Finally, carefully inspect the splines on both shafts and pinions for any signs of wear, hairline cracks or breaking down of the hardened surface finish. If any one of these defects is apparent, then the offending component must be renewed. It should be noted that damage and wear rarely occur in a gearbox which has been properly used and correctly lubricated, unless very high mileages have been covered.

11  It should be assumed that the gearbox sprocket constitutes

part of the layshaft assembly and should therefore be examined along with the rest of that assembly. Clean the sprocket thoroughly and examine it closely, paying particular attention to the condition of the teeth. The sprocket should be renewed if the teeth are hooked, chipped, broken or badly worn. It is considered bad practice to renew one sprocket on its own; both drive sprockets should be renewed as a pair, preferably with a new final drive chain. If this recommendation is not observed, rapid wear resulting from the running of old and new parts together will necessitate even earlier replacement on the next occasion. Examine the splined centre of the sprocket for signs of wear. If any wear is found, renew the sprocket as slight wear between the sprocket and layshaft will rapidly increase due to the torsional forces involved. Remember that the layshaft will probably wear in unison with the sprocket, it is therefore necessary to carry out a close inspection of the layshaft spline.

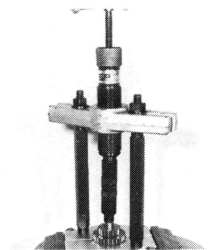

25.5 Use a hydraulic press to remove the mainshaft 2nd gear pinion

25.6 Measure each selector fork to pinion groove clearance

25.9a Both the mainshaft 4th gear pinion ...

25.9b ... and the layshaft 1st gear pinion have a bush fitted within their centres

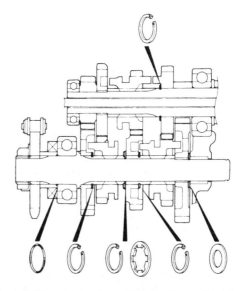

Fig. 1.8 Disposition of gearshaft circlips and washers

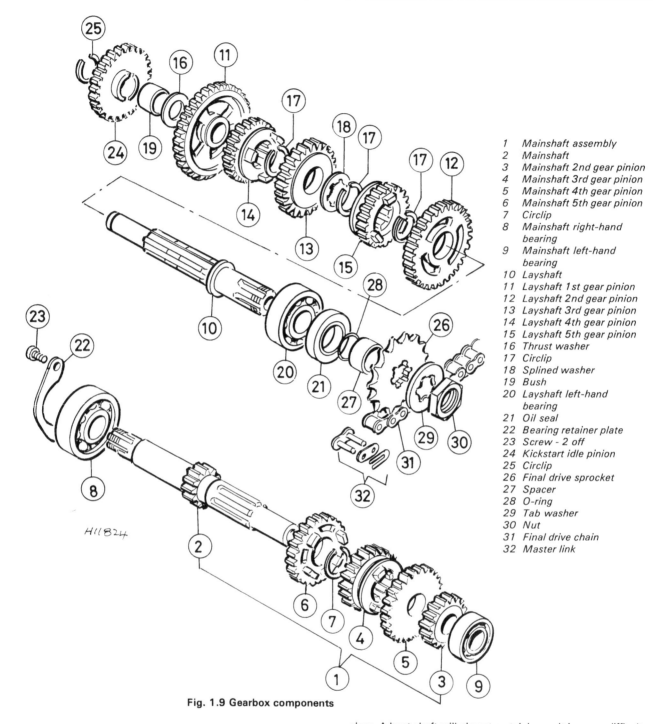

1   Mainshaft assembly
2   Mainshaft
3   Mainshaft 2nd gear pinion
4   Mainshaft 3rd gear pinion
5   Mainshaft 4th gear pinion
6   Mainshaft 5th gear pinion
7   Circlip
8   Mainshaft right-hand
    bearing
9   Mainshaft left-hand
    bearing
10  Layshaft
11  Layshaft 1st gear pinion
12  Layshaft 2nd gear pinion
13  Layshaft 3rd gear pinion
14  Layshaft 4th gear pinion
15  Layshaft 5th gear pinion
16  Thrust washer
17  Circlip
18  Splined washer
19  Bush
20  Layshaft left-hand
    bearing
21  Oil seal
22  Bearing retainer plate
23  Screw - 2 off
24  Kickstart idle pinion
25  Circlip
26  Final drive sprocket
27  Spacer
28  O-ring
29  Tab washer
30  Nut
31  Final drive chain
32  Master link

Fig. 1.9 Gearbox components

## 26  Examination and renovation: gearchange shaft and pawl mechanisms

1   Unless the gearchange shaft and pawl mechanism have been abused in any way or the assemblies have suffered from lack of lubrication, the chances of either component malfunctioning are relatively slight until the machine has been in use for a considerable amount of time.

2   Commence by checking the gearchange shaft for straightness. This may be done by laying the shaft on a sheet of plate glass and attempting to slide a feeler gauge between it and the glass. A bent shaft will almost certainly result in some difficulty in changing gear. Check the condition of the shaft return spring; if it is broken or is showing obvious signs of fatigue, then it must be renewed. Note that the legs of the spring must be placed one each side of the locating pin cast in the shaft endplate. If the spacer placed between the spring and shaft is beginning to show signs of wear or damage, it too should be renewed.

3   Inspect the teeth on the gearchange shaft endplate for damage or wear and look closely for hairline cracks around the base of the teeth. Carry out a similar inspection on the teeth of the pawl mechanism housing. These components will have worn in unison and should therefore be renewed as a matched pair.

4    Prepare a clean piece of rag or paper on a work surface. Dismantle the pawl mechanism, placing each component part on the clean surface as it is removed. Take care when doing this to ensure that each part if positioned so as to obviate any chance of its being incorrectly refitted or transposed with its opposite number.

5    Closely inspect each pawl operating spring for signs of fatigue or failure. These springs must be renewed as a pair, as must both the pawl operating pins and the pawls themselves if either set of component is found to be worn or damaged.

6    Inspect the groove in the pawl housing where it comes into contact with both the pawl lifter and cam guide plates. If the contact surface is seen to be scored or badly worn, seek further advice from an official Suzuki service agent as to whether the component should be renewed. A similar inspection should be carried out on the corresponding contact surfaces of the two plates.

7    Reassemble the renovated pawl mechanism assembly noting that the pawls are handed and must be positioned with the spring recesses offset towards the inside face of the pawl housing as shown in the accompanying photographs; re-wrap it in a rubber band and place it in safe storage until required for engine unit reassembly.

## 27 Examination and renovation: clutch assembly and primary drive

1    Lay the various component parts of the clutch assembly out on a clean work surface, carefully cleaning each item before doing so. The plain and friction plates should all be given a throrough wash in a petrol/paraffin mix to remove all traces of friction material debris and oil sludge; follow this by cleaning the inside of the clutch drum in a similar manner. If this action is not taken, a gradual build-up of contamination will occur and eventually affect the clutch action.

2    Provided the clutch has been reasonably treated and run in the correct type of oil, the bonded linings of the friction plates should last for a considerable time. The obvious sign of the linings having worn beyond their service limit is the advent of clutch slip. To check the degree of wear present on the friction plates, measure the thickness of each plate across the faces of the bonded linings and compare the measurement obtained with the information given in the Specifications Section of this Chapter. If the linings have worn beyond the service limit of 2.6 mm (0.10 in), then the plate must be renewed.

26.2 Check the positioning of the gearchange shaft return spring

26.4 Dismantle and examine the gearchange pawl mechanism ...

26.5 ... withdraw each spring for examination ...

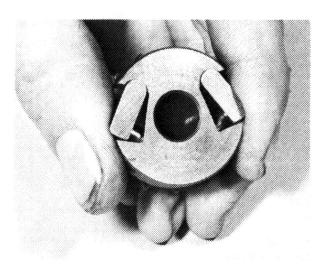

26.7 ... and reassemble the renovated mechanism ready for refitting

3   Check the condition of the tongues around the periphery of each friction plate, at the same time checking the slots in the clutch drum wall. In an extreme case, clutch chatter may have caused the plate tongues to make indentations in the slots; these indentations will trap the plates as they are freed, thereby impairing clutch action. If the damage found is only slight, the indentations can be removed by careful work with a fine file and any burrs removed from the plate tongues in a similar fashion. More extensive damage will necessitate renewal of the parts concerned. Note that there is a definite limit to the amount of material that can be removed from each plate tongue, the minimum tongue width allowable being 11.3 mm (0.45 in).

4   Carry out a check on both the friction and plain plates for any signs of warpage. This can be achieved by laying each plate on a completely flat surface, such as a sheet of plate glass, and attempting to pass a feeler gauge between the plate and the surface. Both the plate and the surface must be cleaned of all contamination. The maximum allowable warpage for each friction plate is 0.4 mm (0.016 in), whereas the maximum allowable warpage for each plain plate is 0.1 mm (0.004 in).

5   The plain plates should be free from scoring and any signs of overheating, which will be apparent in the form of blueing. Check the condition of both the tongues in the inner periphery of each plain plate and the slots of the clutch centre. Any slight damage found on either of these components should be removed by using a method similar to that described for the friction plates and clutch drum. The final check for the plain plates is to measure each plate for thickness. Suzuki give the

standard thickness of each plain plate as 1.6 mm (0.063 in).

6   Inspect the clutch pressure plate for both cracking and signs of overheating. Hairline cracks are most likely to occur around the central bearing hole in the plate and each one of the seven holes that accommodate the springs anchor pins. Check the pressure plate for warpage as if it were a plain or friction plate. Suzuki give no warpage limit for this component, so it is advisable to seek expert advice from an official Suzuki service agent if any warpage found seems excessive.

7   Check the condition of the small central thrust bearing, the pushrod end piece over which it fits and the pushrod(s). Any wear in the bearing will be indicated by roughness felt in the rollers as they are rotated and by cracking of the bearing cage. The pushrod end piece should be renewed if its bearing surfaces show signs of scoring or excessive wear and the two lengths of pushrod should be checked for straightness by rolling them on a flat surface or comparing them with a straight-edge.

8   Measure the free length of each clutch spring. If any one spring has taken a permanent set to a length more than 33.6 mm (1.32 in), then the complete set of springs must be renewed. Check each spring for excessive wear at its contact point with its anchor pin bearing in mind that the two components will have worn in unison. Refit the serviceable springs into the clutch centre, noting that they must be screwed through the casting from its rear face as attempting to insert them from the front will only result in the springs locking in their threads due to their opening action.

9   Closely examine both of the thrust washers for any signs of

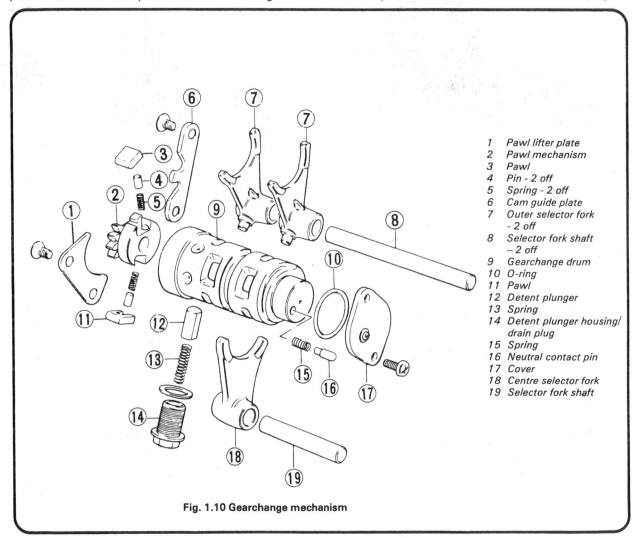

1   Pawl lifter plate
2   Pawl mechanism
3   Pawl
4   Pin - 2 off
5   Spring - 2 off
6   Cam guide plate
7   Outer selector fork
    - 2 off
8   Selector fork shaft
    – 2 off
9   Gearchange drum
10  O-ring
11  Pawl
12  Detent plunger
13  Spring
14  Detent plunger housing/
    drain plug
15  Spring
16  Neutral contact pin
17  Cover
18  Centre selector fork
19  Selector fork shaft

Fig. 1.10 Gearchange mechanism

scoring or overheating and renew each one as necessary. The clutch operating/adjustment mechanism will have remained in the forward section of the left-hand crankcase cover and should not normally require any attention other than periodic lubrication and examination for wear. Pay special attention to the condition of the lever return spring in this mechanism and renew it if it shows signs of fatigue or failure. Access to the mechanism, for the purposes of lubrication and inspection of the internal components, may be gained by removing the two crosshead screws that retain the metal cover plate to the crankcase cover and withdrawing the plate. Inspect the condition of the seal contained within this plate and renew it if it shows signs of damage or deterioration.

10  Primary drive is by means of a crankshaft-mounted pinion driving the clutch by way of a toothed damper unit which is riveted to the rear face of the clutch drum. Because of its method of attachment to the clutch drum, this damper unit is effectively a permanent part of the drum and cannot easily be dismantled for either examination or servicing. Prior to dismantling, check first with a Suzuki dealer whether the damper and rivets are available separately from the clutch drum. Examine both sets of pinion teeth for signs of excessive wear or damage. Note that both sets of teeth will have worn in unison and should therefore be renewed as a matched pair.

11  Note the condition of the Woodruff key which serves to retain the drive pinion to the crankshaft end. If this key shows any signs of wear or damage, then it must be renewed. Wear or damage on the key will invariably mean that the keyways in the pinion and crankshaft will be similarly affected. Apart from renewing these components, the only satisfactory answer to the problem of worn or damaged keyways is either to have them recut to an acceptably larger size and have a new key fitted or to have each keyway refilled by either a welding or metal spraying process and then recut to accommodate a key of the original size. Either one of these alternatives requires a high degree of skill in the use of specialist equipment and the work should, therefore, be placed in the hands of a light engineering company which specialises in such a task. It must be realised that fitting a new key into worn keyways will only, at the best, effect a temporary cure and the turning forces imposed on the crankshaft and pinion will soon reduce any new key to the state of the one removed whilst at the same time wearing the keyways to such a degree that any reclamation or repair may well be impossible.

12  Finally, note the details given in the following Section of this Chapter on the examination of the kickstart driven pinion, which is like the primary driven pinion, effectively part of the clutch drum.

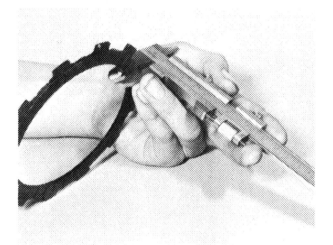

27.2 Measure the amount of wear on each friction plate

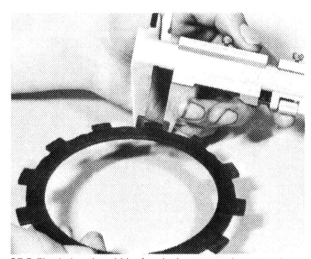

27.3 Check that the width of each plate tongue is no less than the minimum allowed

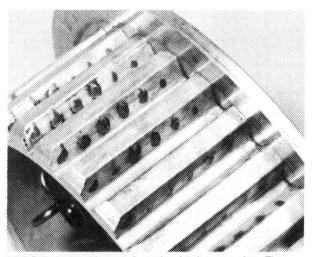

27.5 Remove any damage from the clutch centre slots. The component shown is in good condition, needing no attention

27.7 Inspect the clutch thrust bearing

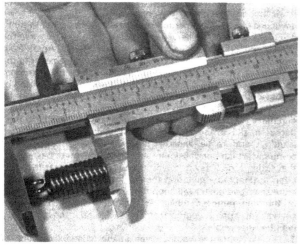

27.8a Measure the free length of each clutch spring

27.8b Refit each spring into the clutch centre from the rear

27.9 Inspect and lubricate the clutch operating mechanism

27.10 Examine the teeth of the clutch damper unit pinion

## 28 Examination and renovation: kickstart assembly

1   Assemble the various component parts that comprise the kickstart assembly and lay them out on a clean work surface ready for inspection. Thoroughly clean and examine each individual component whilst paying particular attention to the following points.

2   If any slipping has been encountered when attempting to turn the engine by means of the kickstart lever, a worn ratchet or drive pinion will invariably be traced as the cause. In the case of the ratchet being damaged or worn, both halves of the assembly must be renewed as a pair; that is, the pinion and the ratchet stop. If the teeth of the pinion are worn, then they must be compared with the condition of the teeth on both the idle pinion and the driven pinion, which is part of the clutch drum assembly. This is because all three pinions will wear in unison and should therefore be renewed as a set. It will be appreciated that any great degree of wear is unlikely to occur between these pinions until the machine has been in use for a considerable amount of time, assuming that the mechanism has always been correctly lubricated and has not been subject to misuse. Any damaged pinion teeth will of course necessitate renewal of the pinion concerned.

3   Inspect the pinion to shaft bearing surface of both the drive and idle pinions for obvious wear or damage. If either shaft is seen to be stepped or is, for example. deeply scored, then the component will have to be renewed. Each pinion has, however, a bushed centre, thereby obviating any need for pinion renewal should its bearing surface be worn or damaged. Suzuki supply a bush ready made to fit the idle pinion but only supply the bush fitted within the drive pinion as part of the pinion assembly. This means that any replacement bush for the drive pinion will need to be made up by a competent motor engineer or a light engineering company who are willing to do such a small job.

4   Removal and fitting of the above mentioned bushes is really a job for the Suzuki service agent or engineering company concerned. This is because each bush if of a very thin-walled construction and will distort very easily when being fitted unless proper shouldered tools are used to press it into position. It may well also be necessary to ream out the centre of each bush in order to obtain the correct fit between shaft and pinion.

5   Examine the kickstart shaft for wear or damage, paying particular attention to the condition of its splined sections. Examine the central splined section in conjunction with the ratchet stop and splined washer that fit over it, checking for obvious wear of the splines which will be indicated by a breaking down of the hardened surfaces. Carry out a similar examination on the shaft end and kickstart lever.

6   It will appear obvious if either of the two springs within the assembly are either broken or fatigued. These, along with any other obviously worn or damaged components should be renewed as a matter of course.

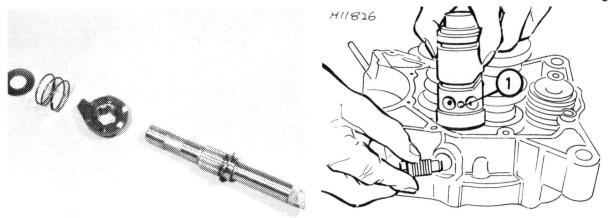

H11826

**Fig. 1.11 Correct position of the detent plunger and gearchange drum**

1    Neutral indent

28.1 Inspect the component parts of the kickstart assembly

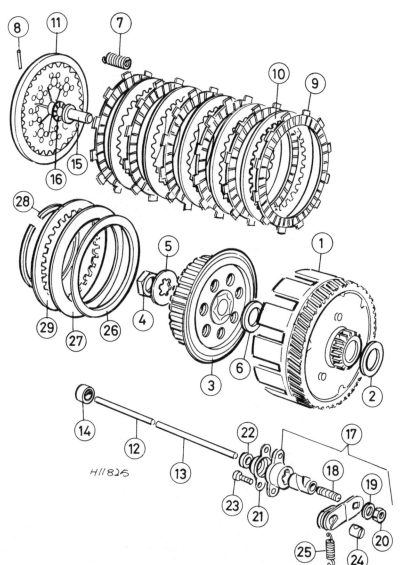

H11826

1    Drum
2    Thrust washer
3    Hub
4    Nut
5    Lock washer
6    Thrust washer
7    Spring - 7 off
8    Anchor pin - 7 off
9    Friction plate – 6 off
     GP100 C and all 125 models,
     5 off GP100 N, X, D models
10   Plain plate – 5 off GP100 C
     and all 125 models, 4 off
     GP100 N, X, D models
11   Pressure plate
12   Right-hand pushrod*
13   Left-hand pushrod*
14   Oil seal
15   Pushrod end piece
16   Thrust bearing
17   Operating mechanism
18   Adjusting screw
19   Washer
20   Locknut
21   Backplate
22   Dust seal
23   Screw - 2 off
24   Cable trunnion
25   Return spring
26   Wave washer seat    } Optional
27   Wave washer           on X and D
28   Retaining ring        models
29   Plain plate

*one piece pushrod on
GP100 C models

**Fig. 1.12 Clutch**

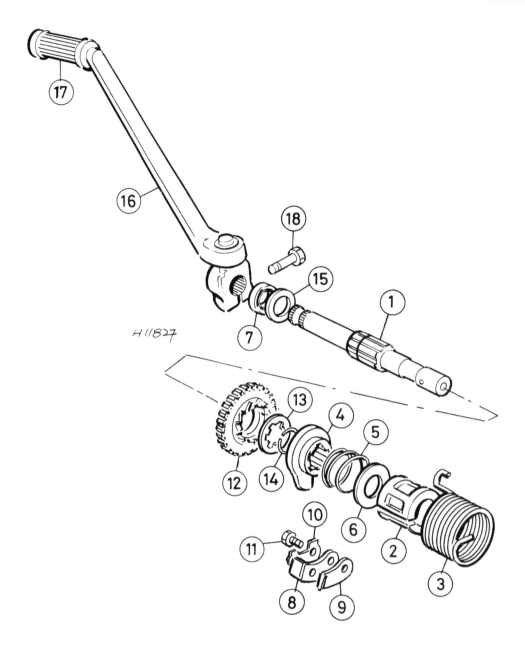

H 11827

**Fig. 1.13 Kickstart mechanism**

| | | |
|---|---|---|
| 1 Kickstart shaft | 7 Oil seal | 12 Drive pinion |
| 2 Spring guide | 8 Ratchet stop retaining | 13 Splined washer |
| 3 Return spring | plate guide | 14 Circlip |
| 4 Ratchet | 9 Ratchet stop retaining | 15 Washer |
| 5 Spring | plate | 16 Kickstart lever |
| 6 Washer | 10 Tab washer | 17 Kickstart rubber |
| | 11 Bolt - 2 off | 18 Pinch bolt |

---

### 29 Examination and renovation: oil pump assembly

1   Because no replacement parts are obtainable for the oil pump fitted to the machines covered in this Manual, the unit is therefore effectively sealed. The only maintenance that can be done is to ensure that the unit body is thoroughly cleaned and given a good, close inspection for any hairline cracks that may be apparent in the parts of the pump body which are subject to stress, that is, around the securing screw holes, pipe union, etc.

2   In normal circumstances, the pump unit can be expected to give long service whilst requiring no maintenance. Do not omit to fit a new gasket between the pump body and crankcase mating surfaces when refitting the pump and always fit new sealing washers when reconnecting the pipe union. Note that

this pipe union is not fitted to the GP100 D, GP125 D and later versions of the GP125 X. The oil feed pipe on these models is a push fit onto the oil pump stub. The type of oil feed pipe fitting will be obvious upon inspection of the pump itself.

3   Wear or damage in either the oil pump drive pinion or its shaft will be obvious on inspection, necessitating the immediate renewal of the component concerned. The shaft pin should also be carefully examined.

## 30 Examination and renovation: tachometer drive assembly

1   With the complete tachometer drive assembly removed from the crankcase, separate the component parts and place them on a clean work surface ready for inspection. As the assembly operates under ideal conditions, aided by the fact that the shaft rotates within a self-lubricating nylon sleeve and its teeth mesh with the nylon drive pinion, it is unlikely that any degree of wear will be observed on any of the component parts until the engine has been running for a considerable amount of time. Once wear is apparent, the part concerned must be renewed.

2   Check the condition of the O-ring fitted to the nylon sleeve. If the ring is damaged or has in any way deteriorated, then it must be renewed; this should be done as a matter of course when rebuilding the complete engine/gearbox unit.

## 31 Examination and renovation: disc valve assembly

1   Reference should be made to the appropriate Section in Chapter 2 for full details of the examination and renovation procedures listed for the disc valve assembly.

## 32 Reassembling the engine/gearbox unit: general

1   Before reassembly of the engine/gearbox unit is commenced, the various component parts should be cleaned thoroughly and placed on a sheet of clean paper, close to the working area.

2   Make sure all traces of old gaskets have been removed and that the mating surfaces are clean and undamaged. One of the best ways to remove old gasket cement is to apply a rag soaked in methylated spirit. This acts as a solvent and will ensure that the cement is removed without resorting to scraping and the consequent risk of damage. If a gasket becomes bonded to the surface through the effects of heat and age, a new sharp razor blade can be used to effect removal. Old gasket compound can also be removed using a soft brass wire brush of the type used for cleaning suede shoes. A considerable amount of scrubbing can take place without fear of damaging the mating surfaces.

3   Gather together all the necessary tools and have available two oil cans, one filled with clean engine oil and one filled with clean gearbox oil. Make sure that all new gaskets and oil seals are to hand, also all replacement parts required. Nothing is more frustrating than having to stop in the middle of a reassembly sequence because a vital gasket or replacement has been overlooked.

4   Make sure that the reassembly area is clean and that there is adequate working space. Many of the smaller bolts are easily sheared if over-tightened. Always use the correct size screwdriver bit for the crosshead screws and never an ordinary screwdriver or punch. If the existing screws show evidence of maltreatment in the past, it is advisable to renew them as a complete set.

5   If the purchase of a replacement set of screws is being contemplated, it is worthwhile considering a set of socket or Allen screws. These are invariably much more robust than the originals, and can be obtained in sets for most machines, in either black or nickel plated finishes. The manufacturers of these screw sets advertise regularly in the motorcycle press.

## 33 Reassembling the engine/gearbox unit: reassembling the gearbox components

1   Having examined and renewed the gearbox components as necessary, the gear clusters can be built up and assembled as a complete unit ready for fitting into the right-hand crankcase half.

2   Refer to the line drawing accompanying this text and proceed to assemble both the mainshaft and layshaft components in the exact order shown whilst noting the following points.

3   Each component must be carefully inspected for any signs of contamination by dirt or grit during assembly and, if necessary, cleaned before being liberally coated with clean gearbox oil on its mating surfaces. When seating each circlip in its groove, take care to ensure that its ends are correctly located in relation to the splines of the shaft; that is, the gap between the circlip ends should fall directly in line with the base of any one of the channels in the splined shaft, as indicated in the various photographs accompanying this text.

4   Upon refitting the 2nd gear pinion to the mainshaft, thoroughly degrease both the surfaces of the shaft and the pinion where they contact each other. Coat the surface of the pinion with Suzuki Lock Super 103Q (Part No 99000-32030) or equivalent and, using a suitable socket or length of thick-walled tube in conjunction with a hammer, drift the pinion onto the shaft. Ensure, whilst doing this, that the threaded end of the shaft is protected from damage by placing it on a wooden surface and that the pinion is kept square to the shaft. Using a micrometer or vernier gauge, keep a close and accurate observation on the distance between the outer faces of the pinion being fitted and the 1st gear pinion which forms a permanent part of the mainshaft (see accompanying photograph). The pinion is correctly fitted when this distance is seen to be between 80.3 – 80.4 mm (3.16 – 3.17 in).

5   Place the assembled gearbox components on a clean piece of card or rag and cover them to protect any ingress of dirt or grit into the assembly during the time taken to prepare the crankcase for their installation.

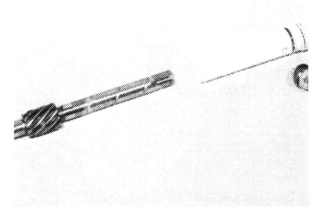

30.1 Inspect the component parts of the tachometer drive assembly

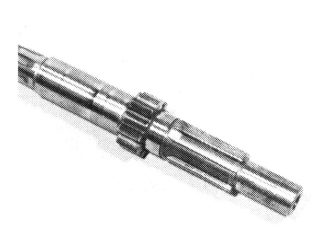

33.4a The gearbox mainshaft

33.4b Fit the mainshaft 5th gear pinion with its circlip

33.4c Fit the mainshaft 3rd gear pinion ...

33.4d ... followed by the 4th gear pinion

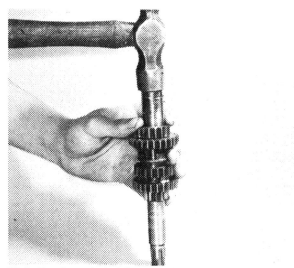

33.4e Tap the mainshaft 2nd gear pinion onto the shaft ...

33.4f ... keeping a close observation on the distance between the 1st and 2nd gear pinions

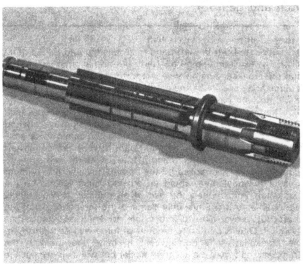

33.4g The gearbox layshaft

33.4h Fit the layshaft 2nd gear pinion with its circlip ...

33.4i ... followed by the 5th gear pinion, the circlip and the thrust washer ...

33.4j ... the 3rd gear pinion with its circlip ...

33.4k ... the 4th gear pinion ...

33.4l ... and finally, the 1st gear pinion and the thrust washer

**34 Reassembling the engine/gearbox unit: fitting the crankcase components and joining the crankcase halves**

1    Place the right-hand crankcase half on a clean area of work surface along with the two oil cans. Proceed to fit the kickstart shaft assembly by first ensuring that the ratchet stop retaining plate securing bolts are properly tightened and locked in position with their tab washer. Slide the splined thrust washer onto the kickstart shaft and secure it in position with the circlip. Note the punch mark at the end of one of the shaft spline channels; this mark must align with the corresponding mark on the edge of the ratchet stop boss once the stop is slid into position. Slide the spring over the boss of the ratchet stop and follow this with the thick plate washer. Note that although these various component parts will have been cleaned during the examination procedure, care must be taken to prevent any further contamination during assembly; do not omit to apply lubricant between any bearing surfaces as each part is fitted.
2    Insert the assembled kickstart shaft into the crankcase half and locate the ratchet stop between its retaining plate and the crankcase casing. Position the return spring end over the locating boss in the crankcase and, using a pair of long-nose pliers, grip the inner end of the spring and rotate it through approximately 90° in a clockwise direction until it can be inserted into the shaft hole. In practice, it was found that this job was made far easier if an assistant was employed to steady the crankcase half so that the two hands were free to grip the pliers. With the return spring properly located, push the plastic guide into position between the shaft and spring.
3    Position the crankcase half on the work surface so that the end of the kickstart shaft is raised slightly clear of the surface; this will preclude any risk of the shaft being pushed out of position. Block the crankcase half so that it is absolutely steady. If the mainshaft bearing retainer plate has been removed, it should now be refitted. It is a good idea to degrease the threads of the plate retaining screws and coat them with a thread locking compound before insertion into the crankcase, as there is no other means provided for ensuring that neither one of these screws will drop out of position whilst the engine is being run.
4    Cup the assembled gear clusters together in both hands, ensuring that the gears are correctly meshed, and lower the complete assembly into the crankcase half. Pay particular attention to the thrust washer, which should remain held in position over the layshaft end until the shaft begins to enter the casing.
5    Fit the gearchange drum into position and rotate it until the neutral indent in the drum end is aligned with the hole for the detent plunger assembly in the crankcase. Refit and tighten the detent plunger assembly, ensuring that the sealing washer remains beneath the housing head. Refit the selector forks into their previously noted positions, ensuring that their fork ends are located correctly into their respective pinion grooves and that their pins are located correctly into the channel of the gearchange drum. Slide the fork shafts into position.
6    The right-hand crankcase half is now complete and should be left in position ready for fitting to the left-hand assembly. Carry out a final lubrication of any components left dry and cover the assembly with a clean cloth to prevent any ingress of dirt. Set the left-hand crankcase half up on wooden blocks on a clean work surface so that once the crankshaft is fitted, its end will remain clear of the surface.
7    Although, on the particular machine dismantled for the writing of this Manual, the crankshaft remained with the left-hand crankcase half as the crankcase halves were separated, this may not always prove to be the case. Indeed, because Suzuki advise that the crankshaft should be drifted into the right-hand crankcase before the two crankcase halves are joined, it would seem logical to suppose that it is purely a matter of chance as to which crankcase half will retain the crankshaft. In practice, it seemed sensible to take the course of least resistance when fitting the crankshaft and to drift it back into the left-hand crankcase half.

8    Lubricate both the left-hand main bearing and the lip of its oil seal with clean engine oil. Carry out a final check to ensure that the crankcase half is well supported around the area of the main bearing and insert the crankshaft into the bearing as far as it will go with hand pressure.
9    The method used to drift the crankshaft into position was to place a long socket or length of thick-walled tube over its end and to tap sharply the end of the socket or tube with a soft-faced hammer whilst taking great care to keep the crankshaft square to the crankcase. Using a tube as a drift will serve to spread the shock imposed by the hammer evenly over a larger area of flywheel then would an ordinary shaft placed on the crankshaft end. Indeed the use of an ordinary drift is highly inadvisable as it will not only damage the threaded end of the shaft but prove almost impossible to keep square to the flywheel. Only a moderate amount of force was needed to tap the crankshaft fully home; bear in mind that heating the crankcase half may make fitting easier but will also dry out the bearings and affect the oil seal. If it is thought that excessive force is being required to fit the crankshaft and that there is a serious risk of the crankshaft or crankcase casting becoming damaged, then it is recommended that aid and advice be sought from an official Suzuki service agent.
10    Firmly push both of the two large locating dowels into their locations in the crankcase mating surface. Lubricate the right-hand main bearing with clean engine oil and thoroughly degrease the mating surface of each crankcase half. Apply a bead of sealing compound (Suzuki Bond No 4 or equivalent) to one of the mating surfaces, taking care not to omit any area of the surface. Raise the left-hand crankcase half from the work surface and lower it onto the supported right-hand crankcase half, taking care to guide the crankshaft and gearbox shaft ends into their respective locations. Push the crankcase halves together with hand pressure as far as they will go whilst noting that the two locating dowels are correctly aligned with their locations in the mating surface. It will now be necessary to tap the left-hand crankcase half with a soft-faced hammer in order to bring the mating surfaces together. On no account should excessive force be used when joining the crankcase halves.
11    Insert all twelve crankcase retaining screws into their previously noted locations, from the left-hand side of the crankcase. Use an impact driver to tighten these screws whilst working in a diagonal sequence so as to obviate any risk of the crankcase halves becoming distorted.
12    With all the crankcase retaining screws fully tightened, wipe away any excess sealing compound from around the mating surfaces and check the free running and operation of the crankshaft and the gearbox components. Any tightness or malfunction will necessitate separation of the crankcases so that the problem may be located and rectified. When satisfied with the operation of the crankcase components, support the crankcase on the work surface so that its right-hand side is facing uppermost ready to receive the gearchange components.

**35 Reassembling the engine/gearbox unit: fitting the gearchange shaft and pawl mechanism**

1    Remove the rubber band from around the assembled pawl mechanism and insert it into the gearchange drum end. Place both the lifter plate and guide plate in position and fit and tighten their securing screws. Note that all four of these screws must have their threads degreased and coated with a thread locking compound before insertion.
2    Lightly grease the length of the gearchange shaft and check that the legs of its spring are placed one each side of the endplate pin. Insert the shaft into its crankcase location, taking care to align its teeth with those of the pawl (as shown in the accompanying photograph) before pushing it fully home. Failure to do this will mean, at the very most, an inefficient gearchange operation.

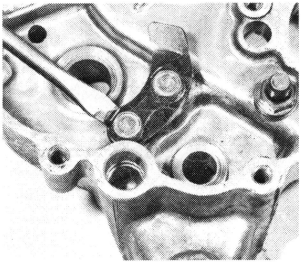

34.1a Lock the kickstart ratchet stop retaining plate bolts in position

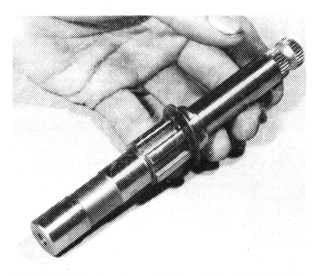

34.1b Fit the thrust washer and circlip onto the kickstart shaft

34.1c Align the mark on the ratchet stop boss with that on the shaft

34.1d Slide the spring into position followed by the thick plate washer

34.2a Locate the ratchet stop between the plate and crankcase

34.2b Locate the kickstart return spring ...

34.2c ... and insert the plastic guide into position

34.4 Lower the mainshaft and layshaft assemblies into the crankcase half

34.5a Position the gearchange drum and fit the detent plunger assembly

34.5b Fit the selector forks into their previously noted positions ...

34.5c ... and slide the fork shafts into position

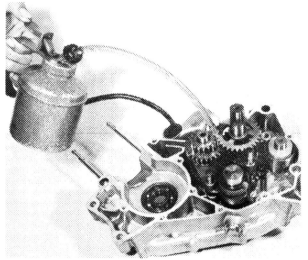

34.6 Lubricate all the component parts

34.9 Use a tubular length of metal to drift the crankshaft into position

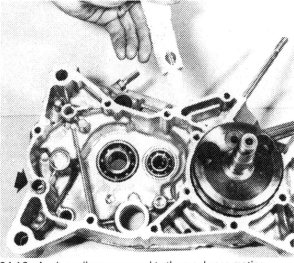

34.10a Apply sealing compound to the crankcase mating surface – arrows indicate locating dowels ...

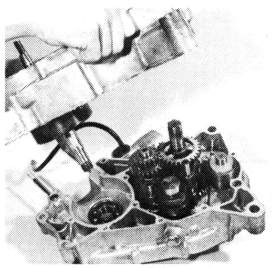

34.10b ... and fit the crankcase halves together

35.1a Insert the gearchange pawl mechanism into the drum end

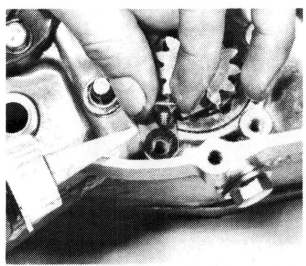

35.1b Apply a thread locking compound to the threads of each screw ...

35.1c ... before fitting the lifter plate and guide plate

35.2 Align the teeth of the shaft with those of the pawl

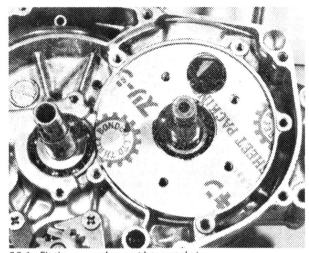

36.1a Fit the new valve seat base gasket ...

### 36 Reassembling the engine/gearbox unit: fitting the disc valve assembly

1   Align the hole in the new valve seat base gasket with the corresponding inlet hole in the crankcase and push the gasket into position. Place the valve seat over the gasket, ensuring that its hole is also in alignment with that in the crankcase. Degrease the threads of each one of the five seat retaining screws and coat them with a thread locking compound before inserting each screw into position. Tighten the screws evenly and in a diagonal sequence.

2   The central splined boss should now be refitted over the crankshaft end; this must be fitted with the finely ground section uppermost in order to avoid damage occurring to the lip of the seal in the valve outer cover. Before pushing the boss into position, thoroughly clean and degrease the edge of the boss that will come into contact with the crankshaft main bearing and coat it with sealing compound (Suzuki Bond No 4 or equivalent). Align the keyway cut in the bore of the boss with that cut for the primary drive pinion key in the crankshaft end and push the boss into position.

3   It is a good idea at this point to fit the primary drive pinion key into its keyway in the crankshaft. This is because some manoeuvring of the splined boss may be necessary in order to allow insertion of the end of the key between it and the crankshaft and this will not be possible once the valve outer cover is fitted.

4   Thinly smear each side of the valve plate with clean engine oil and place the plate over the splines of the central boss. Remember to ensure that the mark made before removal on the outward facing surface of the plate is showing and that the notch in the plate boss is in direct alignment with the keyway in the crankshaft. Failure to do this will result in incorrect valve plate timing which will prevent the engine from starting or from running correctly.

5   Prepare the valve outer cover for fitting to the crankcase. The two O-rings and the crankshaft oil seal will have been renewed during the examination and renovation procedure. Smear the lip of the oil seal with clean engine oil and generously lubricate the valve seat area of the cover. Ensure that the large O-ring is correctly located in its retaining groove and carefully ease the oil seal lip over the splined boss before aligning the cover and pushing it into position. Fit and tighten the five cover retaining screws.

36.1b ... and secure the valve seat in position

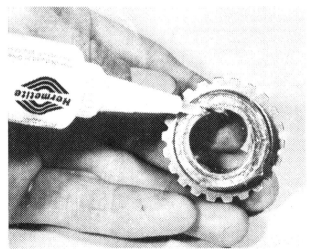

36.2 Coat the inner edge of the disc valve boss with sealing compound ...

36.3 ... before fitting it, together with the Woodruff key

36.5 Secure the disc valve outer cover in position

## 37 Reassembling the engine/gearbox unit: fitting the kickstart drive and idle pinions and the oil pump drive assembly

1   Lubricate the bush of the kickstart idle pinion with clean engine oil and place it in position over the layshaft end. Note that the boss of the pinion must face the crankcase. Fit the pinion retaining circlip and check that it is properly located in its retaining groove.

2   Lubricate the bush of the kickstart drive pinion with clean engine oil and slide it over the kickstart shaft so that its teeth mesh with those of the idle pinion. Place the thrust washer in position on top of the pinion.

3   The oil pump drive assembly may now be fitted by lightly lubricating its shaft with clean engine oil before carefully inserting it into its crankcase location. Check that the teeth of the pump drive pinion mesh correctly with those of the kickstart drive pinion and that the shaft pin is correctly located.

## 38 Reassembling the engine/gearbox unit: fitting the tachometer drive gear

1   Lightly grease the length and teeth of the tachometer drive shaft before inserting it into its nylon sleeve. Check that the new O-ring is correctly located in its retaining groove within the sleeve and slide the complete assembly into the crankcase, pushing it fully home. Fit and tighten the single retaining screw and plain washer. Note that it may be necessary to rotate the shaft during insertion so that its teeth mesh correctly with those of the nylon drive pinion.

## 39 Reassembling the engine/gearbox unit: fitting the primary drive pinion, clutch assembly and right-hand crankcase cover

1   Lock the crankshaft in position by using the method described for removal of the primary drive pinion. Slide the

37.1a The boss of the kickstart idle pinion must face the crankcase

37.1b Secure the pinion in position with the circlip

38.1a Slide the complete tachometer drive assembly into the crankcase ...

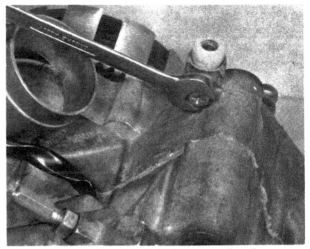

38.1b ... and secure it with the single screw and washer

primary drive pinion into position over the crankshaft end so that the Woodruff key enters the slot in the pinion and gently tap the upper surface of the pinion to seat it on the end of the splined boss of the disc valve assembly. Place a new lock washer over the pinion and fit the retaining nut. This nut should be tightened to a torque loading of 3.6 – 5.0 kgf m (26.0 – 36.0 lbf ft) and tab of the lock washer bent against one of the flats of the nut to lock the nut in position.

2    Place the thrust washer over the end of the mainshaft and slide it down onto the mainshaft bearing. Lubricate both the washer and bearing with clean engine oil and place the clutch drum in position over the mainshaft. Ensure that the teeth of the primary drive pinion mesh with those of the drum by rotating both pinion and drum as the drum is pushed into position. Fit the second of the two thrust washers onto the mainshaft.

3    Before fitting the clutch hub, check that none of the clutch spring ends are protruding above the base of the hub. Any protruding springs will come into contact with the face of the clutch drum thereby causing severe clutch drag and damage to the drum casting. Place the clutch hub in position over the mainshaft, fit a new lock washer and fit and tighten the hub retaining nut to a torque loading of 2.0 – 3.0 kgf m (14.5 – 21.5 lbf ft) whilst holding the mainshaft in position by employing the method used for removal of the nut. Bend the lock washer against one of the flats of the nut to lock the nut in position.

4    Fit the clutch plates into the drum starting with a friction plate followed by a plain plate; building up the layers of plates and finishing with a friction plate. Lubricate the clutch pushrod(s) with clean engine oil and insert into the mainshaft. GP100 C models have a single pushrod and all other models have two short pushrods. Follow this by inserting the pushrod end piece which should also be lubricated. Lightly grease the clutch thrust bearing and place it over the boss of the end piece.

5    Place the pressure plate in position so that the pushrod end piece protrudes through the hole in its centre and the alignment mark on the periphery of the plate aligns with the boss cast in the wall of the clutch hub. Using the special spring tensioning tool mentioned in paragraph 3 of Section 11, pull each spring end out through the pressure plate and insert its anchor pin into position. Take great care to ensure that these pins locate properly into the recesses provided in the plate whilst gripping each one securely with a pair of long-nose pliers so as to obviate any chance of its being dropped into the clutch assembly.

6    Carry out a final check to ensure that all the components contained within the right-hand crankcase cover have been correctly assembled and, where necessary, locked in position. Wipe clean the mating surfaces of both the crankcase and the cover and insert the two locating dowels into position in the

crankcase. Smear the splined end of the kickstart shaft and the lip of the cover oil seal with grease; this will lessen the risk of the seal being damaged as the shaft passes through it. Place a new gasket over the mating surface and push the cover into position over the dowels, tapping it lightly with a soft-faced hammer to seat it properly. Insert each retaining screw into its previously noted position in the cover, tightening it finger-tight. With all the screws in position, proceed to tighten them fully, working in a diagonal sequence to prevent the cover from becoming distorted. Finally, refit the kickstart lever to its shaft end, ensuring that it is correctly positioned and that its retaining bolt is fully tightened.

**40  Reassembling the engine/gearbox unit: fitting the fly-wheel generator assembly**

1    Position the flywheel generator stator plate on the crankcase and fit its three retaining screws and washers, finger-tight. Route the electrical leads from the stator through their retaining grommets and position the end of the loom on the top of the crankcase. Take care to ensure that the single lead to the neutral indicator switch is correctly routed and that none of the leads are allowed to pass over the crankcase to cover mating surfaces.

2    Rotate the stator plate until the alignment marks made on removal align exactly and then lock the plate in position by fully tightening the three retaining screws. If these marks were not made or the stator being fitted is a replacement item, note the index line cast in the plate adjacent to one of the retaining screw dots and align this line with the centre of the appropriate screw before locking the plate in position. The contact breaker gap and ignition timing should be checked as a matter of course before the engine is started. The relevant details will be found in Chapter 3. Make these checks before fitting the rotor for the final time, using locking fluid and torque wrench as described in the following paragraph.

3    Clean and degrease both the taper of the crankshaft and the bore of the flywheel generator rotor where the two components come into contact. Insert the Woodruff key into the crankshaft keyway and push the rotor onto the crankshaft. Gently tap the centre of the rotor with a soft-faced hammer to seat it on the crankshaft taper and then fit the plain washer followed by the spring washer. Clean and degrease the threads of the rotor retaining nut and of the crankshaft end. Apply a thread locking compound to these threads and fit and tighten the nut, finger-tight. Lock the crankshaft in position by employing the method used for rotor removal and tighten the rotor retaining nut to a torque loading of 3.0 – 4.0 kgf m (21.5 – 29.0 lbf ft).

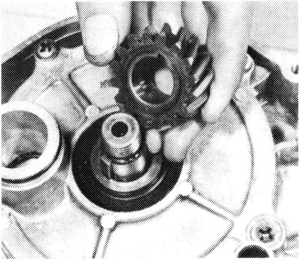

39.1a Fit the primary drive pinion over the crankshaft end ...

39.1b ... tighten the pinion retaining nut ...

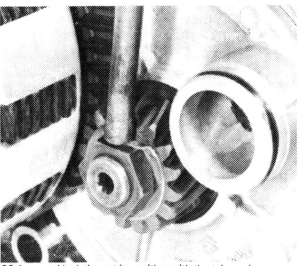

39.1c ... and lock the nut in position with the tab washer

39.2a Place the thrust washer over the mainshaft end ...

39.2b ... followed by the clutch drum and the second thrust washer

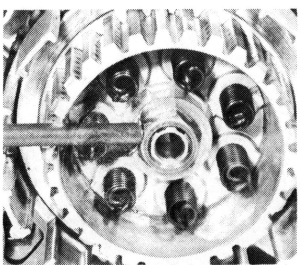

39.3 Lock the clutch drum retaining nut in position with the lock washer

39.4a Insert the shorter length of pushrod into the gearbox mainshaft ...

39.4b ... followed by the pushrod end piece ...

39.4c ... and the greased thrust bearing

39.5 The mark on the periphery of the pressure plate must align with the boss in the clutch hub

39.6a Fit a new gasket over the crankcase mating surface – arrows indicate locating dowels ...

39.6b ... and fit the crankcase cover

40.1 Position the flywheel generator stator plate on the crankcase

40.3a Fit the flywheel generator rotor, followed by the plain washer ...

40.3b ... followed by a serviceable spring washer

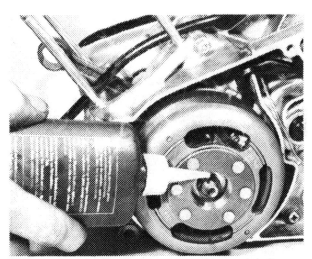

40.3c Apply a thread locking compound to the crankshaft threads ...

40.3d ... before tightening the rotor retaining nut

---

**41 Reassembling the engine/gearbox unit: fitting the gearbox sprocket and neutral indicator switch assembly**

1   Grease the splined end of the layshaft over which the gearbox sprocket fits. This will lessen the risk of the O-ring being damaged as it passes over the threaded end and splines of the shaft. Lightly grease the lip of the shaft oil seal and the surface of the sprocket spacer.

2   Carefully ease the O-ring over the shaft end and push it as far up the shaft splines as possible. Fit the spacer over the shaft and use its end to push the O-ring through the oil seal. With the spacer pushed fully home, wipe any excess grease from the thread of the shaft and fit the gearbox sprocket followed by a new lock washer and the retaining nut. Lock the layshaft in position by using the method described for sprocket removal and tighten the nut to a torque loading of 4.0 – 6.0 kgf m (29.0

– 43.5 lbf ft). Lock the nut in position by bending the lock washer against one of its flats.

3   Remove the component parts of the neutral indicator switch from their storage and lay them out on a clean area of work surface. Where a ball bearing forms part of the switch assembly, grip it between the jaw ends of a pair of long-nose pliers and carefully insert it into the hole provided in the end of the gearchange selector drum. Insert the spring into the drum hole so that it locates on top of the ball and insert the contact pin so that its stepped end locates correctly into the end of the spring. Check that the new O-ring is located correctly on the inside face of the switch cover and fit the cover into position over the drum end so that its contact screw faces the front of the engine unit. Secure the cover in position by fitting and tightening its two retaining screws. Finally, connect the single electrical lead to the switch contact, making sure that the lead is correctly routed before finally tightening the contact screw.

41.2a Ease the O-ring over the layshaft end ...

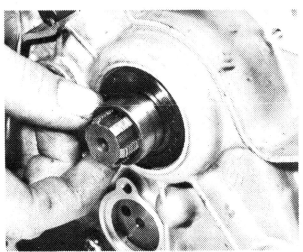

41.2b ... and push it through the oil seal with the spacer

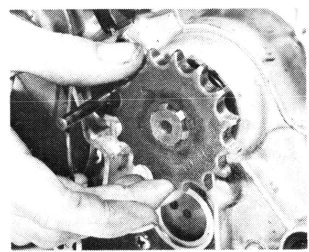

41.2c Fit the gearbox sprocket ...

41.2d ... tighten the sprocket retaining nut ...

41.2e ... and lock the nut in position with the lock washer

41.3a Insert the ball bearing into the gearchange selector drum

41.3b ... followed by the spring ...

41.3c ... and the contact pin

41.3d Connect the electrical lead to the switch contact

## 42 Reassembling the engine/gearbox unit: fitting the oil pump assembly

1   Clean both the oil pump and crankcase mating surfaces. Place a new gasket onto the pump mating surface and align the central driven spigot of the pump with the slot in the end of the drive shaft before fitting the pump into its crankcase housing. With the pump properly seated, fit and tighten its two retaining screws. Note that each of these screws must have a serviceable spring washer located beneath its head.

2   If the oil feed and delivery pipes were detached from the pump during removal of the assembly, they should now be unplugged and reconnected. Ensure that both pipes are a good push fit on their respective stubs and that the larger pipe is correctly retained by its spring clip. Route both pipes through their original locations, taking care to ensure that they are neither twisted nor crimped between any engine components.

## 43 Reassembling the engine/gearbox unit: fitting the small-end bearing, piston, cylinder barrel and cylinder head

1   Position the engine/gearbox unit so that it is upright on the work surface. Rotate the crankshaft to raise the connecting rod to its highest point and thoroughly lubricate the big-end bearing with clean engine oil. Wipe clean the crankcase to cylinder barrel mating surface before easing the new base gasket into position over the barrel retaining studs and packing the crankcase mouth with clean rag in order to prevent any component parts from falling into the crankcase during the following fitting procedures.

2   Lubricate the small-end eye of the connecting rod and the small-end bearing itself with clean engine oil before pushing the bearing into position. Place the piston over the connecting rod so that the arrow cast in the piston crown faces forward and slide the gudgeon pin into position. The pin should be a light sliding fit but if it proves to be tight, warm the piston in hot water to expand the metal around the gudgeon pin bosses. Use new circlips to retain the gudgeon pin, and double check to ensure that each is correctly located in the piston boss groove. If a circlip works loose, it will cause serious engine damage. The circlips should be fitted so that the gap between the circlip ends is well away from the cutout to the side of the gudgeon pin hole. Finally, check that the piston rings have not been disturbed from the positions quoted in Section 24 of this Chapter.

3   Lubricate both the cylinder bore and piston rings with clean engine oil. Position two blocks of wood across the crankcase mouth, one each side of the connecting rod, and carefully lower the piston onto the blocks. This will provide positive support to the piston whilst easing the rings into the bore.

4   Place the cylinder barrel in position over its retaining studs and proceed to lower it carefully down over the piston. Guide the piston crown into the bore and push in on each side of the piston rings so that they slide into the bore. There is a generous lead in one base of the bore which will aid this operation. Take care that the ring ends stay each side of the ring pegs; if the rings ride up over the pegs breakage is certain. With the rings safely inserted into the cylinder bore, remove the blocks from underneath the piston and the rag from the crankcase mouth. Push the cylinder barrel firmly down onto the crankcase and tap it lightly around its upper surface with a soft-faced hammer to ensure that it is properly seated.

5   Clean the mating surfaces of both the cylinder barrel and cylinder head. Carefully ease a new cylinder head gasket over the retaining studs and press it into position on the barrel. Fit the cylinder head and fit and tighten its four retaining nuts (with washers, where necessary), finger-tight. These nuts should now be tightened to a torque loading of 2.3 – 2.7 kgf m (16.5 – 19.5 lbf ft) whilst tightening in even increments and working in a diagonal sequence. This method of tightening will prevent the cylinder head from becoming distorted.

6   It should be noted that on no account should any form of joining compound be applied to the surfaces of either the cylinder head gasket or cylinder barrel base gasket.

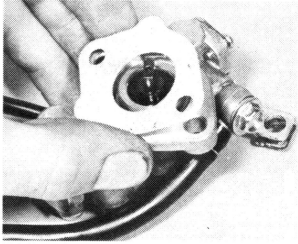

42.1 Always fit a new gasket on the oil pump mating surface

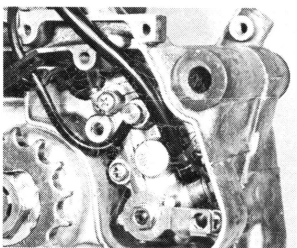

42.2 Reconnect the oil feed and delivery pipes

43.1a Thoroughly lubricate the big-end bearing

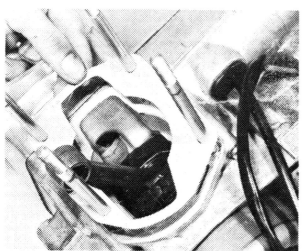

43.1b Fit a new cylinder barrel base gasket

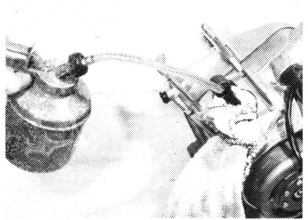

43.1c Pack the crankcase mouth with rag and lubricate the small-end eye

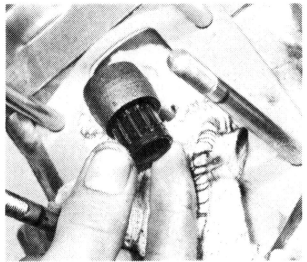

43.2a Fit the small-end bearing

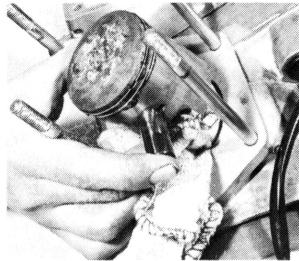

43.2b Fit the piston and retain it with the gudgeon pin

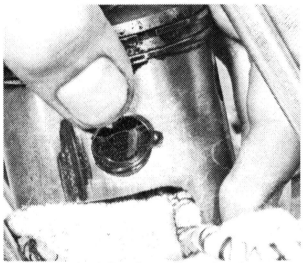

43.2c Position the circlip end gap well away from the cutout

43.4 Carefully lower the cylinder barrel over the piston

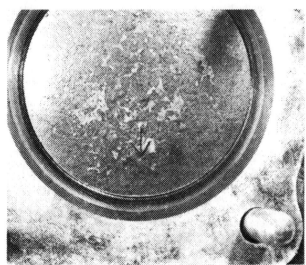

43.5a Check that the arrow on the piston crown is pointing forward ...

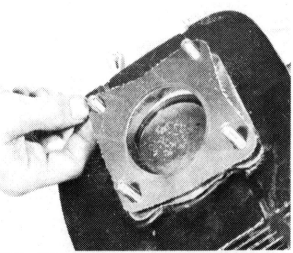

43.5b ... before fitting the new cylinder head gasket, ...

43.5c ... positioning the cylinder head and tightening its retaining nuts

## 44 Fitting the engine/gearbox unit into the frame

1   It is well worth checking at this stage that no component part has been omitted during the various rebuilding sequences. It is better to discover any left-over items now rather than just before the engine is to be started.

2   Fitting of the engine/gearbox unit into the frame is, generally speaking, a direct reversal of the removal procedure. Ideally, the help of one person will be needed in the initial stages of fitting. If this help is not available, then it will be necessary to arrange a stout wooden box or similar item beneath the frame in order to support the engine unit when the mounting bolts are fitted.

3   Check around the frame to ensure that nothing will impede the progress of the engine unit whilst it is being lifted into position. Ease the unit into the frame from the right-hand side of the machine and, once it is aligned with its mounting points, insert from the left-hand side the upper of the two rear mounting bolts. The engine unit may now be pivoted around this bolt so that the two front mounting bolts followed by the one remaining rear mounting bolt can be inserted, all from the left-hand side. Ensure that the appropriate washers have been fitted to each of these bolts before fitting the retaining nut and tightening them to the torque loading figure given in the Specifications Section of this Chapter for the particular bolt size.

4   Check the condition and gap of the spark plug before fitting it to the cylinder head. Do not omit to check that the aluminium crush washer is still attached to the plug; this washer ensures an effective seal between the plug and head casting and serves to keep the plug electrodes the required height from the piston crown. Ideally, a new spark plug should be fitted after a full engine rebuild. Applying a smear of graphite grease to the threads of the plug will greatly lessen the chances of the plug becoming seized in the cylinder head during engine use.

5   The carburettor may now be removed from its position on the frame top tubes and refitted to its location in the right-hand crankcase cover. Ensure that the mouth of the carburettor is inserted correctly into its mounting stub before the retaining clamp is tightened. If this mounting stub was found to be damaged or split and hence renewed, insertion of the carburettor mouth may prove to be difficult due to the tightness of the stub. This problem may be alleviated by smearing the lip of the stub with a solution of soapy water.

6   With the carburettor secured in position, reconnect each of the five pipes to the various points on its body whilst using the sketch made during the removal procedure for reference. The photographs accompanying this text will give an indication of

the routing of these pipes. Note the pipe guide fitted beneath the head of one of the float chamber retaining screws and the spring clip which retains the fuel feed pipe in position. It was found in practice, that because access to the innermost of these pipes was severely limited, the use of a pair of long-nose pliers was invaluable when easing the pipes over their stubs. To avoid damaging the pipes however, the jaws of the pliers must be padded with tape. It is now necessary to delay the fitting of the carburettor cover and cable seal until both a check for fuel leaks and a check on oil pump adjustment have been carried out. It is, necessary however, to ensure that the throttle cable is correctly adjusted and that both it and the choke cable function smoothly over their full operating range.

7   Adjustment of the throttle cable is correct when there is 1.0 – 1.5 mm (0.04 – 0.06 in) of free movement in the cable outer when it is pulled out of the adjuster at the handlebar end. If this adjustment is found to be incorrect, loosen the adjuster locknut and rotate the adjuster the required amount before retightening the locknut.

8   Reconnect the tachometer drive cable to its location at the top of the gearbox housing. Ensure that the end of the cable inner is correctly located in the drive gear assembly before fitting and tightening the knurled retaining ring.

9   Move to the left-hand side of the machine and lubricate the remaining length of clutch pushrod with clean engine oil before inserting it through its crankcase seal and into the centre of the gearbox mainshaft.

10  Insert the oil pump control cable through its adjuster at the crankcase end and push up on the end of the pump lever to allow insertion of the cable nipple into its nylon holder. The oil pump must now be checked for correct adjustment in accordance with the following procedure.

11  Rotate the throttle twistgrip until the circular indicator mark on the base of the throttle slide comes into alignment with the upper edge of the carburettor mouth (see Figure 2.5). With the throttle set in this position, check that the mark scribed on the pump lever boss is in exact alignment with the mark cast in the pump body. If this is not the case, then the marks should be made to align by rotating the control cable adjuster, after having released its locknut. On completion of the adjustment procedure, retighten the locknut whilst holding the cable adjuster in position and slide the rubber sealing cap down the cable to cover the adjuster. It should be noted that any adjustment of the oil pump control cable may well affect the adjustment of the throttle cable. It is therefore, necessary to check the throttle cable for correct adjustment before proceeding further.

12  Reconnect the oil feed pipe to the stub at the base of the oil tank and retain it in position with the spring clip. Refill the oil tank with oil of the specified type and proceed to check that both the feed pipe and pump are primed. This can be accomplished by loosening the cross-headed screw on the side of the pump body and waiting for a steady stream of oil to emerge before retightening the screw. Failure to carry out this priming procedure will mean that the engine will run dry when first started, with the subequent risk of seizure.

13  Place the oil pump cover plate in position and fit and tighten its two retaining screws. Wipe clean the crankcase to cover mating surfaces and position a new gasket over the forward section of crankcase surface. Place the forward section of crankcase cover in position and insert its four retaining screws. Fit the one short screw in the position noted during the removal procedure. Tighten the screws evenly and in a diagonal sequence in order to lessen the risk of the cover becoming distorted. It should be remembered when placing the cover in position, that the end of the clutch pushrod must enter the adjuster assembly; any malalignment of the adjuster and pushrod will prevent the cover from seating.

14  The clutch should now be adjusted by first loosening the knurled lock ring of the handlebar lever adjuster and rotating the adjuster until it abuts against the lever bracket. Move to the adjuster at the crankcase end of the cable and loosen its locknut. Rotate the adjuster until free play can be felt in the cable inner. It is now necessary to free the adjuster screw

contained within the end of the clutch operating lever by releasing its locknut. Turn this screw inwards until its end is felt to contact the clutch pushrod and then turn it outwards between $\frac{1}{4}$ and $\frac{1}{2}$ of a turn. Lock the screw in position by retightening its locknut whilst taking care to ensure that the screw is not allowed to move from its set position. Complete the adjustment procedure by rotating the adjuster at the crankcase end of the cable to give 2 – 3 mm (0.08 – 0.12 in) of free play (measured between the pivot end of the handlebar lever and its retaining clamp) in the cable inner. Hold the adjuster in position and tighten its locknut. Any subsequent fine adjustment to the amount of free play in the cable inner may be achieved by rotating the knurled adjuster at the handlebar lever and before locking it in position by tightening its lock ring. Slide the rubber sealing cap into position over the adjuster at the crankcase end of the cable.

15 Note at this point that full details of setting the contact breaker gap and carrying out a static check of the ignition timing are contained within the relevant Sections of Chapter 3. Both of these procedures should have been carried out during generator reassembly, the details of which were given in Section 40.

16 Wipe clean the mating surfaces of the left-hand crankcase cover and the contact breaker/clutch adjuster cover before placing both the new gasket and cover in position. Fit and tighten the four cover retaining screws. Fully tighten the spark plug and reconnect its suppressor cap whilst routing the HT lead so that it cannot chafe on any frame or engine component parts.

17 Reconnect the electrical leads of the flywheel generator stator assembly, ensuring that they are correctly routed and clipped to the frame downtube.

18 Manoeuvre the air filter housing into position so that its hose end is correctly located over the crankcase inlet stub and its mounting bracket is aligned with that of the frame. Do not omit to fit the hose retaining clamp before easing the hose into position. If the hose has been renewed and proves difficult to fit over the crankcase stub, its fitting may be made easier by the application of a solution of soapy water to its inner edge. Tighten the hose retaining clamp and fit and tighten the two frame mounting screws with their washers.

19 The battery should now be refitted to the machine and its leads reconnected. Observe the polarity markings of the battery and ensure that the earth lead is connected to the negative (-) terminal. Both terminal connections must be clean and free from corrosion. To prevent corrosion from occurring, smear both connections with a liberal coating of petroleum jelly; do not use ordinary grease. Take care to route the battery vent pipe correctly so that it is not trapped between any frame components and so that its end is placed well clear of the lower frame tubes.

20 Refitting of the exhaust system is a straightforward process. Place a new sealing ring in the recess provided in the cylinder barrel before manoeuvring the system into position. The spring washers fitted beneath the pipe to barrel retaining bolt heads must be in good condition and the nut of the swinging arm fork pivot shaft should be tightened to a torque loading of 4.5 – 7.0 kgf m (32.5 – 50.5 lbf ft).

21 Loop the final drive chain around the gearbox and rear wheel sprockets and reconnect the two ends with the split link. It is most important that the spring clip of this link is correctly fitted with the closed end facing the normal direction of chain travel. It is quite possible that the rear wheel will have to be moved forward in order to place enough slack in the chain to allow insertion of the link. In either case, the chain must be checked for correct tension and adjusted accordingly by referring to the instructions given in Chapter 5 of this Manual.

22 With the final drive chain refitted and correctly tensioned, place the rear section of the left-hand crankcase cover in position and secure it by fitting and tightening its four retaining screws. The gearchange lever can now be slid into position on its shaft. Ensure that the lever is positioned correctly in relation to the footrest and that its bolt hole is aligned with the channel

in the shaft spline before inserting and tightening its retaining bolt. It was noticed on the model used for this Manual, that the lever retaining bolt had begun to work loose. It was, therefore, considered a good idea to apply a coating of thread locking compound to the threads of the bolt in order to prevent a recurrence of this problem.

23 Remount the fuel tank on the machine. Carry out a check to ensure that no part of the tank is in direct contact with the frame as any metal to metal contact will lead to eventual failure of the tank structure. Reconnect the fuel pipe to the fuel tap and retain it in position with the spring clip. Turn the tap lever to the 'On' position and carefully check both ends of the pipe for any signs of fuel leakage. On no account should fuel be allowed to come into contact with hot engine castings; if this is allowed to happen, fire may result causing serious personal injury.

24 On completion of a satisfactory fuel leak check, wipe clean the mating surface of both the carburettor cover and crankcase cover. Fit a new gasket to the crankcase cover and press both the carburettor cover and cable seal assembly into position so that they interlock correctly. Fit the three retaining screws through the cover and the four retaining screws through the seal retaining plate and tighten them fully. Refit the innermost of the two small blanking plugs into the front of the carburettor housing.

25 Check that the gearbox oil drain plug (detent plunger housing) has been properly tightened. Remove the plastic filler plug from the right-hand crankcase cover and replenish the gearbox with 850 cc (1.50 Imp pint) of SAE 20W/40 oil. Refit and tighten the filler plug.

26 Carry out a final check around the engine unit to ensure that all component parts have been correctly fitted and, where applicable, are functioning correctly. Check all electrical connections for security and check that all bolts, nuts, screws and fasteners have been tightened. Refit the seat and sidepanels.

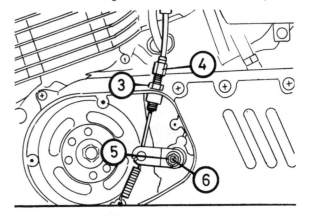

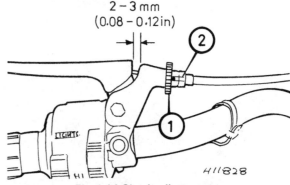

**Fig. 1.14 Clutch adjustment**

| | |
|---|---|
| 1 Knurled locking ring | 4 Adjusting screw |
| 2 Adjusting screw | 5 Locknut |
| 3 Locknut | 6 Adjusting screw |

44.5 The carburettor mounting stub must be in a serviceable condition

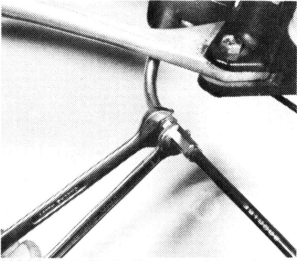
44.7 Adjustment of the throttle cable is carried out at the handlebar end of the cable

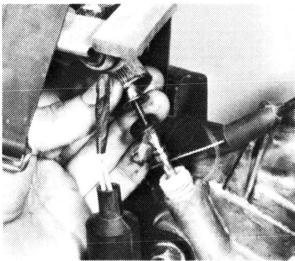
44.8 Reconnect the tachometer drive cable

44.9 Insert the second length of clutch pushrod

44.10a Insert the cable nipple into its nylon holder ...

44.10b ... before fitting the cable holder to the oil pump lever

44.11a Rotate the oil pump control cable adjuster ...

44.11b ... until the marks on the pump lever and pump body align

44.12a Connect the oil feed pipe to the oil tank ...

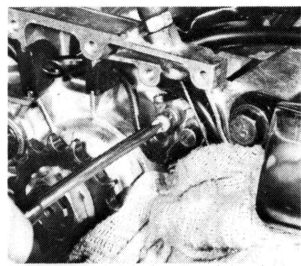

44.12b ... and prime both the pipe and pump by loosening the priming screw

44.13a Refit the oil pump cover plate ...

44.13b ... followed by the forward section of crankcase cover

44.14a Commence clutch adjustment at the clutch lever screw

44.14b Rotate the cable adjuster to obtain the correct amount of cable free play ...

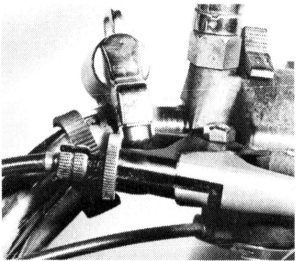

44.14c ... and carry out any subsequent fine adjustment at the handlebar lever

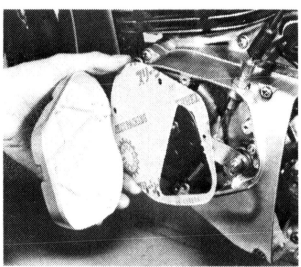

44.16 Place a new gasket beneath the outer cover

44.18a Ensure that both the inlet hose retaining clamp ...

44.18b ... and the air filter housing mounting screws are fully tightened

44.20a Position a new sealing ring in the recess provided in the cylinder barrel

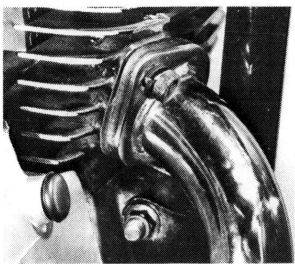

44.20b Place serviceable spring washers beneath the heads of the exhaust pipe retaining bolts

44.20c Tighten the swinging arm pivot nut to the correct torque loading

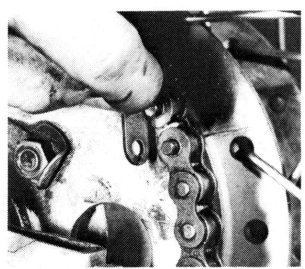

44.21a Push the final drive chain connecting link into position ...

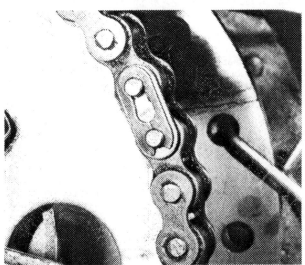

44.21b ... and ensure that its spring clip faces in the right direction

44.23a Push the fuel tank onto its front mounting rubbers ...

44.23b ... and check that the tank rear mounting rubber is correctly positioned

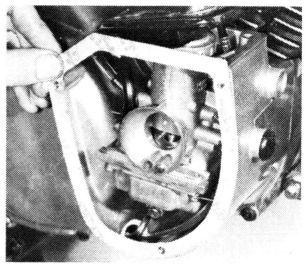

44.24 A serviceable gasket must be fitted beneath the carburettor cover

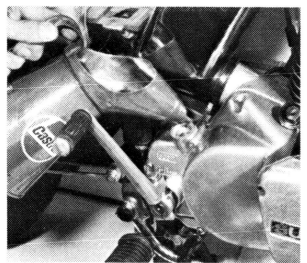

44.25 Replenish the gearbox with oil

### 45 Starting and running the rebuilt engine/gearbox unit

1    Open the petrol tap to allow fuel to flow to the carburettor, close the carburettor choke and start the engine. Raise the choke as soon as the engine will run evenly and keep it running at a low speed for a few moments to permit the oil to circulate through the lubrication system. Do not open the throttle with the choke closed.

2    Bear in mind that the engine parts should be liberally coated with oil during assembly, so the engine will tend to smoke heavily for a few minutes until the excess oil is burnt away. Do not despair if the engine will not fire up at first, as it is quite likely that the excess oil will foul the sparking plug, necessitating its removal and cleaning. When the engine does start, listen carefully for any unusual noises, and if present, establish, and if necessary rectify, the cause. Check around the engine for any signs of oil leakage or blowing gaskets.

3    Make sure each gear engages correctly and that all controls function effectively, particularly the brakes. This is an essential last check before taking the machine on the road.

### 46 Taking the rebuilt machine on the road

1    Any rebuilt engine will need time to settle down, even if parts have been replaced in their original order. For this reason it is highly advisable to treat the machine gently for the first few miles to ensure oil has circulated throughout the lubrication system and that new parts fitted have begun to bed down.

2    Even greater care is necessary if the engine has been rebored or if a new crankshaft has been fitted. In the case of a rebore, the engine will have to be run in again, as if the machine were new. This means greater use of the gearbox and a restraining hand on the throttle until at least 500 miles have been covered. There is no point in keeping to any set speed limit; the main requirement is to keep a light loading on the engine and to gradually work up performance until the 500 mile mark is reached. These recommendations can be lessened to an extent when only a new crankshaft is fitted. Experience is the best guide since it is easy to tell when an engine is running freely. Remember that a good seal between the piston and the cylinder barrel is essential for the correct functioning of the engine. A rebored two-stroke engine will require more careful running-in, over a long period, than its four-stroke counterpart. There is a far greater risk of engine seizure during the first hundred miles if the engine is permitted to work hard.

3    If at any time a lubrication failure is suspected, stop the engine immediately and investigate the cause. If an engine is run without oil, even for a short period, irreparable engine damage is inevitable.

4    Do not on any account add oil to the petrol under the mistaken belief that a little extra oil will improve the engine lubrication. Apart from creating excess smoke, the addition of oil will make the mixture much weaker, with the consequent risk of overheating and engine seizure. The oil pump alone should provide full engine lubrication.

5    Do not tamper with the exhaust system or run the engine without the baffle fitted to the silencer. Unwarranted changes in the exhaust system will have a marked effect on engine performance, invariably for the worse. The same advice applies to dispensing with the air cleaner or the air cleaner element.

6    With the initial run completed, wait until the engine unit has cooled and recheck all fittings and fasteners for security. Readjust any controls that may have settled down during their initial use.

## 47 Fault diagnosis: engine

| Symptom | Cause | Remedy |
|---|---|---|
| Engine will not start | Defective spark plug | Remove the plug and lay it on cylinder head. Check whether sparking occurs when ignition is switched on and engine rotated. |
| | Dirty or closed contact breaker points | Check condition of points and whether gap is correct. |
| | Faulty or disconnected condenser | Check whether points arc when separated. Replace condenser if evidence of arcing. |
| Engine runs unevenly | Ignition and/or fuel system fault | Check each system independently, as though engine will not start. |
| | Blowing cylinder head gasket | Leak should be evident from oil leakage where gas escapes. |
| | Incorrect ignition timing | Check accuracy and if necessary reset. |
| Lack of power | Fault in fuel system or incorrect ignition timing | See above. |
| | Choked silencer | Remove and clean out baffles. |
| | Blocked cylinder barrel exhaust port | Carry out engine decarbonising procedure detailed in Chapter 1. |
| High fuel/oil consumption | Cylinder barrel in need of rebore and o/s piston | Fit new rings and piston after rebore. |
| | Oil leaks or air leaks from damaged gaskets or oil seals | Trace source of leak and replace damaged gasket and/or seal. |
| Excessive mechanical noise | Worn cylinder barrel (piston slap) | Rebore and fit o/s piston. |
| | Worn small end bearings (rattle) | Replace needle roller bearing (caged) and if necessary, gudgeon pin. |
| | Worn big end bearing (knock) | Fit replacement crankshaft assembly. |
| | Worn main bearings (rumble) | Fit new journal bearings and seals. |
| Engine overheats and fades | Pre-ignition and/or weak mixture | Check carburettor settings. Check also whether plug grade is correct. |
| | Lubrication failure | Check oil pump setting and whether oil tank is empty. |

## 48 Fault diagnosis: clutch

| Symptom | Cause | Remedy |
|---|---|---|
| Engine speed increases but machine does not respond | Clutch slip | Check clutch adjustment for free play at handlebar lever. Check condition of clutch friction plate linings. |
| Difficulty in engaging gears. Gear changes jerky and machine creeps forward, even when clutch is withdrawn. Difficulty in selecting neutral | Clutch drag | Check clutch adjustment for too much free play. Check for burrs on clutch plate tongues or indentations in clutch drum slots. Dress with file if damage not too great. |
| | Clutch assembly loose on mainshaft | Check tightness of retaining nut. If loose, fit new tab washer and retighten. |
| Operation action stiff | Damaged, trapped or frayed control cable | Check cable and replace if necessary. Make sure cable is lubricated and has no sharp bends. |
| | Bent pushrod(s) | Renew. |

**49 Fault diagnosis: gearbox**

| Symptom | Cause | Remedy |
| --- | --- | --- |
| Difficulty in engaging gears | Gear selector forks bent<br>Gear cluster assembled incorrectly | Renew.<br>Check that thrust washers are located correctly. |
| Machine jumps out of gear | Worn dogs on ends of gear pinions<br>Selector drum pawls stuck | Renew pinions involved.<br>Free pawl assembly. |
| Gear lever does not return to normal position | Broken return spring | Renew spring. |
| Kickstart does not return when engine is turned over or started | Broken or poorly tensioned return spring | Renew spring or retension. |
| Kickstart slips | Kickstart drive pinion internals, pawls or springs worn badly | Renew all worn parts. |

# Chapter 2 Fuel system and lubrication

## Contents

## Specifications

### Fuel tank

| | |
|---|---|
| Total capacity | 9.8 litres (2.2 Imp gal) |
| Reserve capacity | 2.0 litres (0.4 Imp gal) |

### Fuel grade

Unleaded or low-lead, minimum octane rating 85/90 RON

### Carburettor

| | GP100 | GP125 |
|---|---|---|
| Make | Mikuni | Mikuni |
| Type | VM 22 SS | VM 24 SS |
| ID number: | | |
| C models | 39200 | 39111 |
| N and X models | 39201 | 39111 |
| D and L models | 39240 | 39111 |
| Bore size | 22.0 mm (0.87 in) | 24.0 mm (0.94 in) |
| Main jet: | | |
| C models | 97.5 | 90 |
| N and X models | 97.5 (105 optional) | 90 |
| D and L models | 105 (100 optional) | 90 |
| Needle jet | 0-8 | 0-2 |
| Jet needle | 4P6-2 | 4EJ14-3 |
| Pilot jet: | | |
| C models | 27.5 | 25 |
| N, X, D and L models | 25 | 25 |
| Pilot outlet | 0.6 | 0.8 |
| Pilot air screw setting (turns out from fully in) | $1\frac{1}{2}$ | $1\frac{1}{2}$ |
| By-pass | 1.0 | 1.2 |
| Throttle valve cutaway | 2.5 | 2.5 |
| Float height | $23.5 \pm 1.0$ mm $(0.93 \pm 0.04$ in) | |
| Engine idling speed | $1300 \pm 150$ rpm | |

### Air cleaner

| | |
|---|---|
| Element type | Oiled polyurethane foam |

### Lubrication

| | |
|---|---|
| Engine: | |
| Type | Suzuki CCI – Pump fed total-loss system |
| Oil capacity | 1.2 litres (2.3 Imp pint) |
| Oil type | Suzuki CCI, Suzuki CCI Super or any good quality non-diluent 2-stroke oil |
| Oil pump discharge rate (fully open) | 1.30 – 1.60 cc (0.054 – 0.056 Imp oz) over 2 mins at 2000 rpm |
| Gearbox: | |
| Type | Oil bath |
| Oil capacity: | |
| At oil change | 800 cc (1.41 Imp pint) |
| At engine overhaul | 850 cc (1.50 Imp pint) |
| Oil type | SAE 20W/40 |

## 1  General description

The design and function of the fuel system fitted to the machines covered in this Manual is as follows. Fuel is gravity-fed from the frame-mounted fuel tank to the float chamber of the Mikuni carburettor via a three-position fuel tap which incorporates a small removable filter. Air is drawn into the carburettor via an oil-impregnated polyurethane foam filter element which is contained in a housing mounted just to the rear of the cylinder barrel.

At low engine speeds, the proportions of air and atomised fuel which form the combustion mixture, are controlled by a pilot circuit, these being regulated by a combination of throttle stop and pilot mixture screw settings. As the twistgrip control is turned, the cylindrical throttle valve is lifted, allowing a greater volume of air to be drawn through the carburettor choke. The passage of air across the top of the needle jet causes fuel to be drawn up through the main jet and needle jet by venturi action.

The amount of fuel entering the engine is at this stage metered by the needle jet assembly, in which a tapered needle is drawn upwards with the throttle valve, allowing increasing amounts of fuel to enter the combustion mixture, as the throttle is opened. Eventually, the rate of flow of the fuel is restricted by the main jet, which has been selected to give the correct mixture at maximum throttle opening.

The point of induction on a two-stroke engine is normally controlled by the piston skirt, which covers and uncovers ports machined in the cylinder bore. On the Suzuki models covered in this Manual, a supplementary timing system is employed to enable more efficient induction timing. This device is known as a rotary valve, and consists of a thin steel disc, which is mounted on the crankshaft and enclosed by a lined casing.

This casing has a port machined in it, and this aligns with a similar cutout in the valve disc. At a predetermined crankshaft position, the valve opens, allowing the combustion mixture to be drawn into the crankcase in the usual way. The valve then closes, preventing any back leakage of the mixture. The mixture is then fed to the combustion chamber via transfer ports in the normal manner.

Engine lubrication is by a pump fed system known as Suzuki CCI. Two-stroke oil is gravity fed from a frame-mounted tank to the oil pump. The pump is driven by the engine via the primary and kickstart drive assemblies, and is also interconnected by a Bowden cable to the throttle twistgrip. Thus the amount of oil passed by the pump is varied according to the engine speed and throttle setting.

Oil from the pump is fed directly to the main and big-end bearings, and to the disc valve assembly. The remaining engine components are lubricated by splash. Any residual oil is drawn into the combustion chamber along with the incoming fuel mixture, and is therefore burnt.

Lubrication for the transmission components is catered for by oil contained within the gearbox casing, isolated from the working parts of the engine proper.

## 2  Fuel tank : removal and fitting

1   Prior to removing the fuel tank, the dualseat must be detached by unscrewing the two bolts which retain the seat base to the frame and then lifting it up and rearwards off its forward mounting point.
2   Remove the fuel tank by first turning the fuel tap lever to 'Off' position and releasing the fuel pipe retaining clip. This will allow the pipe to be pulled off the stub at the rear of the tap. Careful use of a small screwdriver may be necessary to help ease the pipe off the stub. Once the pipe is detached, allow any fuel in the pipe to drain into a small clean container. The tank may now be detached from the frame by unscrewing the single retaining bolt at the rear of the tank and pulling the tank up and rearwards off its front mounting rubbers. Place the tank retaining bolt, together with any associated mounting compo-

nents in a safe place ready for refitting. Inspect the mounting rubbers for signs of damage or deterioration and if necessary renew them before refitting of the tank is due to take place.
3   Store the tank in a safe place whilst it is removed from the machine, well away from any naked lights or flames. It will otherwise represent a considerable fire or explosion hazard. Check that the tap is not leaking and that it cannot be accidentally knocked into the 'On' position. It is well worth taking simple precautions to protect the paint finish of the tank whilst in storage. Placing the tank on a soft protected surface and covering it with a protective cloth or mat may well avoid damage being caused to the finish by dirt, grit, dropped tools, etc.
4   To refit the tank, reverse the procedure adopted for its removal. Move it from side to side before it is fully home, so that the rubber buffers engage with the guide channels correctly. If difficulty is encountered in engaging the front of the tank with the rubber buffers, apply a small amount of petrol to the buffers to ease location. Secure the tank with the single retaining bolt whilst ensuring that the mounting components are correctly located and that there is no metal to metal contact between the tank and frame.
5   Finally, always carry out a leak check on the fuel pipe connections after fitting the tank and turning the tap lever to the 'On' position. Any leaks found must be cured; as well as wasting fuel, any petrol dropping onto hot engine castings may well result in a fire or explosion occurring.

## 3  Fuel tap : cleaning the filter element

1   The type of fuel tap fitted to the GP100 and 125 U, C, N and early X models incorporates a fuel filter element located within its base, retained by a plastic bowl and O-ring assembly. The purpose of this filter element is to prevent any contamination in the fuel being passed to the carburettor, thus preventing any one of the jets within the carburettor becoming blocked. Contamination caught by the filter is deposited in the plastic bowl which must be removed, together with the filter element, at frequent intervals for cleaning and examination. The recommended service interval for these components is every 4000 miles (6000 km). It should, however, be realised that any loss in performance or a refusal of the engine to run for any more than a short period of time, might be attributable to fuel starvation caused by a blocked or partially blocked filter element. GP100 and 125 D, L and late X models do not have a fuel filter element fitted and rely entirely on the gauze filter fitted over the filter stack inside the fuel tank. It is, however, worth removing the plastic bowl from time to time to clean away any sediment that may have accumulated, and to check for signs of water or contamination in the fuel supply. If evidence of fuel contamination is found, the filter bowl and filter (where fitted) will require cleaning more frequently than the intervals recommended. It may even be necessary to remove and clean out the fuel tank.
2   To remove the filter element, turn the tap lever to the 'Off' position, place an open-ended spanner over the squared end of the bowl and turn it anti-clockwise to unscrew the bowl. It was found in practice that the O-ring between the bowl and the tap casing had formed a semi-permanent seal between the two components and some effort was required to effect an initial release of the bowl. With the bowl and O-ring removed, the filter element may be gently eased out of its location within the tap whilst taking care to note the positioning of the hole within the filter element in relation to the corresponding fuel line within the tap.
3   Inspect the condition of the O-ring; if it is flattened, perished, or in any way damaged, then it must be renewed. Clean the filter element by rinsing it in clean fuel. Any stubborn traces of contamination may be removed from the element by gently brushing it with a small soft-bristled brush soaked in fuel; a used toothbrush is ideal. The filter bowl may be cleaned by

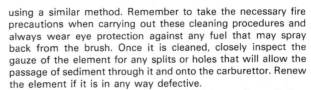

2.4a Engage the fuel tank with its front mounting rubbers

using a similar method. Remember to take the necessary fire precautions when carrying out these cleaning procedures and always wear eye protection against any fuel that may spray back from the brush. Once it is cleaned, closely inspect the gauze of the element for any splits or holes that will allow the passage of sediment through it and onto the carburettor. Renew the element if it is in any way defective.

4    Clean the fuel tap base of any remaining sediment before inserting the element into position, fitting the serviceable O-ring and fitting and tightening the filter bowl. Take care not to overtighten the bowl, it need only be nipped tight. Finally, turn the tap lever to the 'On' position and carry out a check for any leakage of fuel around the bowl to tap joint. Cure any leak found by nipping the bowl a little tighter. If this fails, remove the bowl and check that the O-ring has seated correctly.

### 4    Fuel tap : removal, fitting and curing of leaks

1    It will be found, should the fuel tap spring a leak, that the leak will emit from any one of three points on the tap: the tap to tank joint, the tap lever joint or the filter bowl joint. The latter

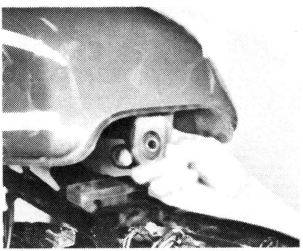

2.4b Locate rear mounting rubber before bolting tank in place

3.2a Remove the bowl from the fuel tap ...

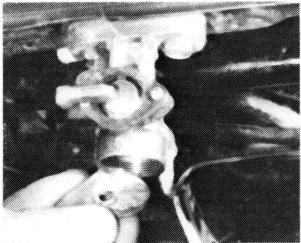

3.2b ... and withdraw the fuel filter element

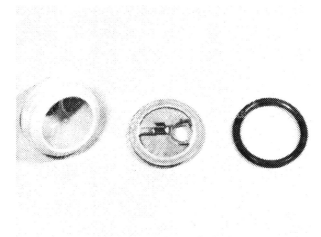

3.3 Inspect and clean the fuel filter component parts

of these points has been fully dealt with in the preceding Section of this Chapter; however, the initial attempt at curing any one of these leaks remains the same, that of simply check-tightening the screws or bolts securing the component concerned. If this simple check fails to effect a cure, then the leak must be stopped by following the relevant procedure laid down in the following paragraphs.

2    Any leak from the tap to tank joint will be caused by the O-ring between the two components being defective. Renewal of the O-ring will call for removal of the tap from the tank. This may be done with the tank in situ but will necessitate draining the tank of as much fuel as possible. This is easily accomplished by detaching the fuel feed pipe from the tap and replacing it with a similar length of pipe which can be routed into a clean container placed next to the machine. Note that this container should be constructed of metal and have a removable top which incorporates an adequate seal; it should also be of an adequate capacity. Before turning the tap lever to the 'On' position, which will allow the fuel to drain, ensure that all necessary fire precautions have been taken; it is most important that the machine is positioned in an area where there is good ventilation.

3    With the tank thus drained of fuel, remove the tap by unscrewing the two retaining bolts with their sealing washers. Take care when drawing the tap unit out of the tank, not to damage either of the filter stacks by letting them come into contact with the edge of the tank orifice. Place the tap unit on a clean work surface and ease the O-ring out of its retaining groove in the tap body. If the filter stacks show signs of being contaminated, then they should now be detached for cleaning. Note that the filter stack may be of either one piece construction covering both outlets or two separate filter stacks. The method of cleaning, however, is the same for each type. Clean the stack(s) by rinsing in clean fuel. Any stubborn traces of contamination may be removed by gently brushing the stack with a soft-bristled brush which has been soaked in fuel; a used toothbrush is ideal. Remember to take the necessary fire precautions when carrying out this cleaning procedure and always wear eye protection against any fuel that may spray back from the brush. On completion of cleaning, closely inspect the gauze area of each stack for any splits or holes that will allow the passage of sediment through it and into the fuel tap. Renew the stack if it is in any way defective.

4    Clean the O-ring retaining groove and the inner area of the tap before replacing each stack and pressing the new O-ring into position. Wipe clean the area around the orifice of the tank where the O-ring makes contact and carefully reinsert the tap into the tank. Fit a new sealing washer to each of the tap retaining bolts and fit and tighten each bolt to secure the tap in position. The fuel feed pipe may now be reconnected to the tap stub and the tank replenished with fuel. Check any disturbed connections for fuel leaks before starting and riding the machine.

5    Any leak from the tap lever joint will be caused by the seal contained between the lever and tap body being defective. Renewal of this seal will call for draining of the tank prior to removal of the tap lever retaining plate. The method best used for draining the tank is described in paragraph 2 of this Section. To remove the lever retaining plate, simply unscrew and remove the two plate retaining screws before pulling the plate, together with the lever, from position. Remove the defective seal from its location within the tap body, clean the seal location of any sediment, fit the new seal into the tap body, relocate the lever and retaining plate and secure the plate with the two retaining screws. Reconnect the fuel feed pipe to the tap stub, replenish the tank with fuel and carry out a check for fuel leaks on all the disturbed connections before starting and riding the machine.

## 5   Fuel feed pipe : examination

1    The fuel feed pipe is made from thin-walled synthetic rubber and is of the push-on type. It is only necessary to replace the pipe if it becomes hard or splits. It is unlikely that the retaining clip should need replacing due to fatigue as the main seal between the pipe and union is effected by an interference fit. Suzuki recommend that the fuel feed pipe be renewed every two years as a matter of course.

## 6   Carburettor : removal and fitting

1    To remove the carburettor, move to the right-hand side of the machine and detach the cable seal retaining plate from the top of the carburettor housing by removing its four retaining screws. Work the seal, together with its retaining plate, up the throttle and choke cables until it is well clear of the engine casing. Inspect the seal whilst doing this; if it shows signs of severe damage or deterioration, then it must be renewed before the engine is restarted.

2    Detach the cover from the carburettor housing by removing its three retaining screws and pulling the cover from position. With the carburettor thus exposed, it will be seen that a total of five pipes are routed through the carburettor housing to connect at various points on the carburettor body. Although a photograph is provided with the text of this Chapter, it is important that the correct fitted position of each one of these pipes is noted for reference during fitting; making a quick sketch is by far the best method of achieving this.

3    Before detaching these pipes from the carburettor, check that the fuel tap lever is turned to the 'Off' position and be prepared for the small amount of fuel that will issue from the end of the fuel feed pipe once it is disconnected. Have a small container or piece of rag handy in which to collect this fuel and take care to observe the necessary fire precautions. Pull each pipe off its retaining stub, releasing any retaining clips where necessary and making use of the flat of a small screwdriver to help ease any stubborn pipe from position.

4    Move the front of the carburettor housing and remove the innermost of the two small blanking plugs. This will enable a screwdriver to be passed through the housing wall in order to slacken the carburettor mounting clamp. Ease the carburettor outwards off its mounting stub and manoeuvre it clear of its housing. The carburettor can now be detached from the throttle and choke assemblies by unscrewing the cap of the mixing chamber and withdrawing the throttle valve assembly and then unscrewing the large hexagonal cap which serves to retain the choke valve assembly in position. Unless specific attention is required, both the throttle and choke valve assemblies may be left to hang by their operating cables. The carburettor body may now be placed on a clean work surface ready for examination.

5    Fitting the carburettor is a straightforward reversal of the removal procedure, whilst noting the following points. When refitting the throttle valve assembly into the carburettor body, take great care to ensure that the jet needle enters the needle jet smoothly and correctly and that the slot cut in the side of the throttle valve is in correct alignment with the tag cast in the carburettor body.

6    Ensure that the mouth of the carburettor is inserted correctly into its mounting stub before the retaining clamp is tightened. If the stub was found to be damaged or split and therefore renewed, insertion of the carburettor mouth may prove to be difficult due to the tightness of the stub. This problem can be alleviated by smearing the lip of the stub with a solution of soapy water.

7    With the carburettor secured in position, refer to the sketch drawn during the removal procedure and reconnect each of the five pipes to the various points of the carburettor body. Take note of the pipe guide fitted beneath the head of one of the float chamber retaining screws and the spring clip which retains the fuel feed pipe in position. It was found in practice, that because access to the innermost of these pipes was severely limited, the use of a pair of long-nose pliers was invaluable when easing the pipes over their respective stubs. To avoid causing damage to the pipe ends, the jaws of these pliers must be padded with

tape. With all the pipes reconnected, turn the fuel tap lever to the 'On' position and carefully check around the carburettor body and pipe connections for any signs of fuel leakage. Any leaks found must be cured before proceeding further; leakage of fuel will not only present a serious fire hazard but may well adversely affect the fuel/air mixture passing into the crankcase.

8    On completion of a satisfactory leak check, return the fuel tap lever to the 'Off' position and check that both the throttle and choke controls function smoothly over their full operating range. Note that adjustment of the throttle cable is correct when there is 1.0 – 1.5 mm (0.04 – 0.06 in) of free movement in the cable outer when it is pulled out of the adjuster at the handlebar end. If this adjustment is found to be incorrect, loosen the adjuster locknut and rotate the adjuster the required amount before retightening the locknut.

9    Finally, wipe clean the mating surface of both the carburettor housing cover and the housing itself. Before refitting the cover, check that its gasket is serviceable; a damaged gasket must be renewed as any leakage of air through this joint will adversely affect the fuel/air mixture passing into the crankcase as well as allowing the ingress of road dirt and water into the carburettor. The cover and cable seal must be seen to have interlocked correctly before their retaining screws are fitted and tightened.

6.1 Detach the cable seal retaining plate ...

6.2a ... followed by the carburettor housing cover ...

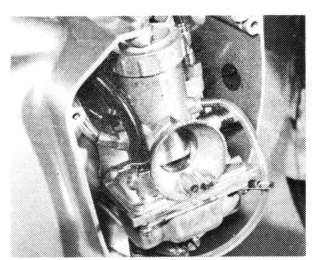

6.2b ... to gain access to the carburettor

6.4 Slacken the carburettor mounting clamp

6.6 The carburettor mounting stub must be in good condition

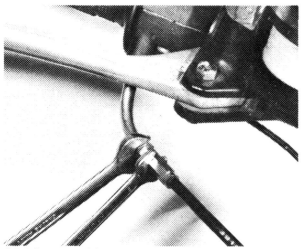

6.8 Adjustment of the throttle cable is carried out at the handlebar end of the cable

6.9 The carburettor cover gasket must be in good condition

## 7   Carburettor : dismantling, examination, renovation and reassembly

1   Before dismantling the carburettor, cover an area of the work surface with clean paper or rag. This will not only prevent any components that are placed upon it from becoming contaminated with dirt, moisture or grit but, by making them more visible, will also prevent the many small components removed from the carburettor body from becoming lost.
2   Proceed to dismantle the carburettor by removing the four screws and spring washers that retain the float chamber to the main body of the carburettor. Note the fitted position of the pipe guide fitted beneath the head of one of these screws. In practice, it was found that the float chamber had become quite firmly adhered to the carburettor body and some gentle persuasion was needed to remove it. Tapping around the joint with a soft-faced hammer may serve to break this seal, otherwise it will be necessary to place the flat of a small screwdriver between the side of the chamber and the lip of the body in order to lever the two components apart. Take great care, when using the latter of these two methods, not to place any great strain on the component castings; the two components should part fairly easily.
3   With the float chamber thus removed and placed to one side, pull the pivot pin from the twin float assembly and lift the floats from position. The float needle can now be displaced from its seating and should be put aside in a safe place for examination at a later stage. It is very small and easily lost if care is not taken to store it in a safe place.
4   Unscrew and remove the single crosshead screw which serves to retain the float needle seat retaining plate in position between the two float pivot pin columns. Withdraw the plate and, using a pair of long-nose pliers, pull the needle seat out of its location in the carburettor body. In order to avoid causing any damage to the seat, it is best to pad the jaws of these pliers with tape.
5   Unscrew and remove the main jet and carefully hook its washer out of its location in the carburettor body. Note that, when unscrewing any jet from the carburettor, a close fitting screwdriver of the correct type must be used to prevent damage occurring to the soft material from which the jet is constructed. With the main jet and its washer removed, the needle jet may be pushed out of its location so that it leaves the carburettor body through the top of the mixing chamber. Take note of the alignment pin cast in the needle jet location for reference when refitting.

6   Unscrew and remove the pilot jet. In practice, it was found that this jet was very tight and great care combined with some effort was needed to free it. Note the setting of the throttle stop screw by counting the number of turns required to screw it fully in. Remove the throttle stop screw, taking care to retain its spring.
7   Note the setting of and remove the pilot air screw with its spring. Failure to note the settings of the aforementioned screws will make it less easy to 'retune' the carburettor after it has been reassembled and refitted to the machine. The main body of the carburettor is now devoid of all removable components and should be placed to one side in readiness for cleaning.
8   The only removable component fitted to the float chamber of the carburettor is the drain plug which takes the form of a single slotted screw with a sealing washer located beneath its head. It is not necessary to remove this plug except for renewal of the sealing washer or replacement of the screw itself.
9   Prior to examination of the carburettor component parts, clean each part thoroughly in clean fuel before placing it on a piece of clean rag or paper. Use a soft nylon-bristled brush to remove any stubborn contamination on the castings and blow dry each part with a jet of compressed air. Avoid using a piece of rag for cleaning since there is always risk of particles of lint obstructing the airways or jet orifices. Never use a piece of wire or any pointed metal object to clear a blocked jet, it is only too easy to enlarge a jet under these circumstances and increase the rate of petrol consumption. If an air line is not available, a blast of air from a tyre pump will usually suffice. If all else fails to clear a blocked jet, remove a bristle from the soft-bristled brush and carefully pass it through the jet to clear the blockage.
10   Check each casting for cracks or damage and check that each mating surface is flat by laying a straight-edge along its length. Any distorted casting must be replaced with a serviceable item.
11   Remove all O-rings and sealing gaskets from the component parts and replace them with new items. Ensure that, where applicable, they are correctly seated in their retaining grooves. The springs on the throttle stop and pilot air screws should now be carefully inspected for signs of corrosion and fatigue and renewed if necessary.
12   The seating area of the float needle will wear after lengthy service and should be closely examined with a magnifying glass. Wear usually takes the form of a ridge or groove, which will cause the float needle to seat imperfectly. If the needle has to be renewed, remember that the needle seat will have worn in unison and in extreme cases, will also need renewing.

13 Closely examine the twin float assembly for signs of damage, especially around the soldered joints. Shake the float to establish if a leak is present. Although it is theoretically possible to repair a brass float by soldering, any attempt to do so is likely to cause a small but dramatic explosion, having a detrimental effect on both the float and the operator. It is far preferable to renew the float, although a safer temporary repair may be made with Araldite or Petseal.

14 Move the machine and inspect the throttle valve for wear. This wear will be denoted by polished areas on the external diameter. Excessive wear will allow air leaks, weakening the mixture, and producing erratic slow running. Many mysterious carburation maladies may be attributed to this defect, the only cure being to renew the valve, and if worn badly in corresponding areas, the carburettor body. If removal of the valve is necessary, grasp the valve firmly in one hand whilst compressing the return spring against the carburettor top with the other. Lift the circular plastic ring (late GP125 models only) out of the throttle valve, disengage the throttle cable from its retaining slot in the valve, and withdraw the retaining plate, followed by the jet needle and its clip. Before renewing the valve, inspect each individual component part, commencing with the needle.

15 Examine the needle carefully for scratches or wear along its length. Ensure that the needle is not bent by rolling it on a flat surface, such as a sheet of plate glass. If in doubt as to the condition of the needle, return it to an official Suzuki service agent who will be able to give further advice and, if necessary, provide a new component.

16 Inspect the return spring for signs of fatigue, failure or severe corrosion and renew it if found necessary. The procedure adopted for reassembly of the throttle valve component parts should be a direct reversal of that used for dismantling.

17 Late GP125 models have a circular plastic ring fitted inside the throttle valve, to restrict the amount of throttle opening. This modification was necessary to comply with the government regulations on 125cc machines introduced in 1982.

18 Inspect the surface of the choke valve for signs of scratches or excessive wear and the return spring for signs of fatigue, failure or corrosion. Renew each component as necessary. Both the valve and spring, along with the hexagonal retaining cap, can be withdrawn from the cable after the cable and nipple has been disengaged from its retaining slot in the valve. This is easily achieved by compressing the spring against the retaining cap in order to allow the valve to be moved up and then sideways to clear the nipple.

19 Prior to reassembly of the carburettor, check that all the component parts, both new and old, are clean and laid out on a piece of clean rag or paper in a logical order. On no account use excessive force when reassembling the carburettor because it is easy to shear a jet or some of the smaller screws. Furthermore, the carburettor is cast in a zinc based alloy which itself does not have a high tensile strength. If any of the castings are damaged during reassembly, they will almost certainly have to be renewed.

20 Reassembly is basically a reversal of the dismantling procedure, whilst noting the following points. If in doubt as to the correct fitted position of a component part, refer either to the figure accompanying this text or to the appropriate photograph. When fitting the throttle stop and pilot air screws, ensure that each screw is first screwed fully in, until it seats lightly and then set to its previously noted position. Alternatively, set the pilot air screw $1\frac{1}{2}$ turns out from fully in; the setting of the throttle stop screw will then have to be determined by following the adjustment procedure listed in Section 8 of this Chapter. Do not omit to align the slot in the top of the needle jet with the alignment pin cast in the jet location before pushing the jet fully home. Finally, smear the external diameter of the throttle valve with light machine oil before refitting the carburettor to the machine.

7.9 Take care whilst cleaning jet orifices or airways

7.11a Renew the float chamber to carburettor body gasket

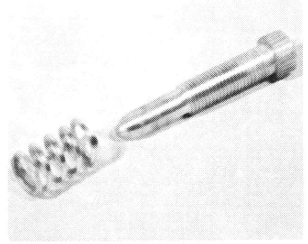

7.11b Inspect the spring on the throttle stop ...

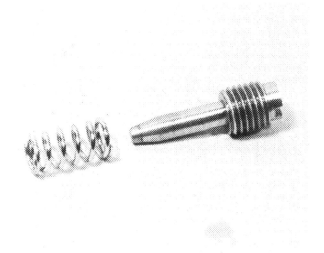

7.11c ... and the spring on the pilot air screw

7.12 Inspect the seating area of the float needle and its seat

7.13 Closely examine the twin float assembly

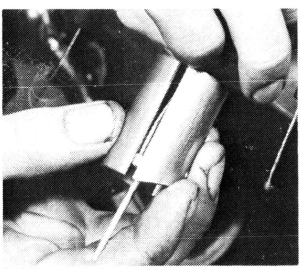

7.14 Disengage the throttle cable from the valve

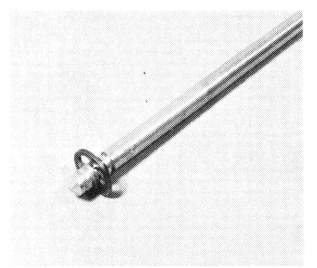

7.16a Check the pressure of the jet needle clip ...

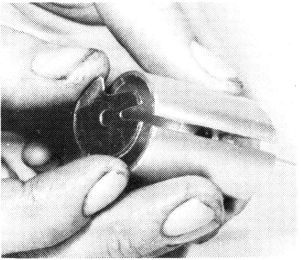

7.16b ... fit the retaining plate over the needle end ...

7.16c ... check the seal within the mixing chamber cap ...

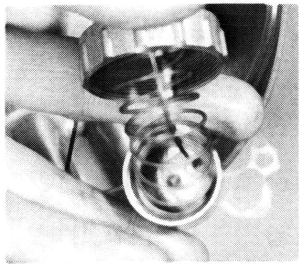

7.16d ... and fit the spring, cap and cable to the throttle valve

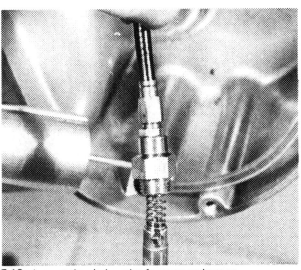

7.18a Inspect the choke valve for wear or damage ...

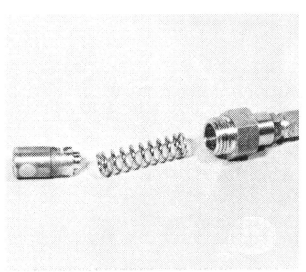

7.18b ... and renew each component part as necessary

7.20a Clean the drain plug and renew its sealing washer ...

7.20b ... before fitting the plug to the float chamber

7.20c Refit the throttle stop screw with its spring ...

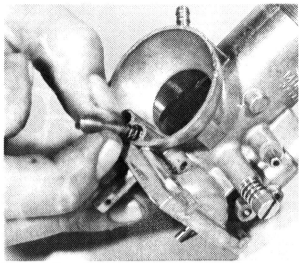

7.20d ... followed by the pilot air screw with its spring

7.20e Clean the needle jet ...

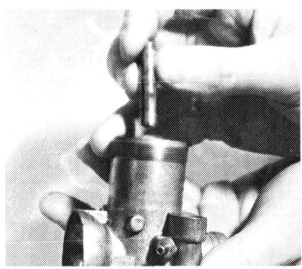

7.20f ... before inserting it into the carburettor body ...

7.20g ... and slotting the end of the jet over the alignment pin

7.20h Clean the main jet and its seating washer ...

7.20i ... fit the washer and screw the main jet into the needle jet

7.20j Push the float needle set into its location ...

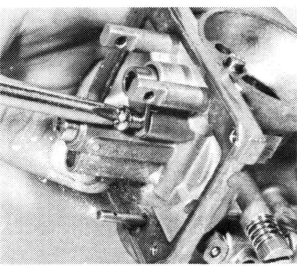

7.20k ... and secure it in position with the retaining plate and screw

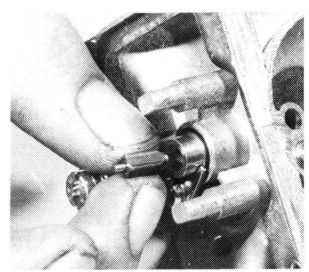

7.20l Fit the float needle into its seat

7.20m Retain the float assembly in position with its pivot pin

7.20n Clean the pilot jet ...

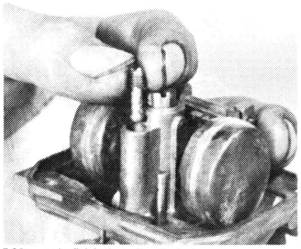

7.20o ... and refit it in the carburettor body

7.20p Refit the float chamber to the carburettor body

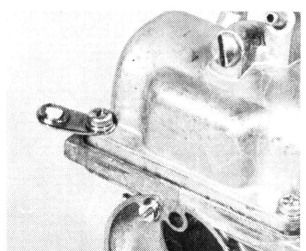

7.20q Note the fitted position of the pipe guide

## 8 Carburettor : adjustment

1 If flooding of the carburettor or excessive mixture weakness have been experienced, it is wise to start operations by checking the float level, which will involve detaching the carburettor, if not already removed, inverting it and removing the float chamber followed by the chamber to carburettor body sealing gasket. Take care to avoid tearing this gasket during removal. In practice, it was found that removal of the twin float assembly provided enough access for the tip of a thin-bladed knife to be carefully inserted beneath the gasket at various points around its inner edge and that gentle levering at these points was enough to free the gasket undamaged.

2 To measure the float level, stand the inverted carburettor on a flat and level work surface and measure the distance between the gasket surface of the carburettor body and the furthest point of the float as shown in the photograph accompanying this text. The float level is correct when the measurement obtained is $23.5 \pm 1.0$ mm ($0.93 \pm 0.04$ in). If necessary, bend the small tongue, sited between the two floats, to obtain the correct setting.

3 With the carburettor correctly refitted, start the engine and allow it to reach its normal operating temperature. Remove the outermost of the two small blanking plugs from the front of the carburettor housing and stop the engine. The carburettor should now be adjusted so that the correct engine idle speed is obtained by first turning the pilot air screw fully in and then unscrewing it $1\frac{1}{2}$ turns. Restart the engine and with the shank of the screwdriver inserted through the orifice in the carburettor housing, turn the throttle stop screw until the engine reaches its slowest possible reliable idle speed. The pilot air screw should now be turned slowly clockwise or anti-clockwise within a range of $\frac{1}{4}$ turn either side of its initial setting until the engine reaches the point where it idles most smoothly. At this point, the engine should be idling at the recommended speed of $1300 \pm 150$ rpm; if the reading on the tachometer indicates an idle speed which is slightly outside the recommended speed, then the throttle stop screw should be turned until the indicated speed is correct. Refit the small blanking plug and relocate the carburettor housing cover.

4 Always guard against the possibility of incorrect carburettor adjustment which will result in a weak mixture. Two-stroke engines are very susceptible to this type of fault, causing rapid overheating and often subsequent engine seizure. Changes in carburation leading to a weak mixture will occur if the air cleaner is removed or disconnected, or if the silencer is tampered with in any way. Above all, do not add oil to the petrol, in the mistaken belief that it will aid lubrication. Adequate lubrication is provided by the throttle controlled oil pump.

8.1 Remove the float chamber sealing gasket ...

8.2 ... to check the float level

## 9 Carburettor : settings

1 The various jet sizes, the throttle valve cutaway and the jet needle position are all settings that are predetermined by the manufacturer. Under normal circumstances, it is unlikely that these settings will require modification, even though there is provision made. If a change appears necessary, it can often be attributed to a developing engine fault. Check with the Specifications Section of this Chapter if there is any doubt as to the size of the component parts fitted. Note that the fitted position of the jet needle clip is indicated by the suffix number of the jet needle identification. For example, on the GP 100 models fitted with a jet needle identification 4P6-2, the number 2 indicates that the fitted position of the clip is in the 2nd groove down from the top of the needle.

2 If the machine is found to perform badly on a particular throttle setting, check that the float level is correct, that all pipes to the carburettor are in sound condition and that the air filter assembly and the exhaust system are both in good order before

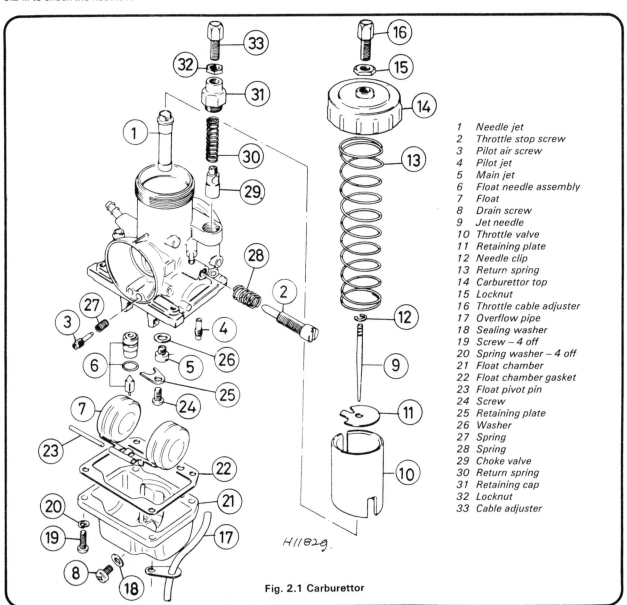

1   Needle jet
2   Throttle stop screw
3   Pilot air screw
4   Pilot jet
5   Main jet
6   Float needle assembly
7   Float
8   Drain screw
9   Jet needle
10  Throttle valve
11  Retaining plate
12  Needle clip
13  Return spring
14  Carburettor top
15  Locknut
16  Throttle cable adjuster
17  Overflow pipe
18  Sealing washer
19  Screw – 4 off
20  Spring washer – 4 off
21  Float chamber
22  Float chamber gasket
23  Float pivot pin
24  Screw
25  Retaining plate
26  Washer
27  Spring
28  Spring
29  Choke valve
30  Return spring
31  Retaining cap
32  Locknut
33  Cable adjuster

H11829.

Fig. 2.1 Carburettor

riding the machine on the throttle setting concerned for a distance of approximately 6 miles.

3   On completion of the ride, remove the spark plug and inspect the insulator and electrodes for condition and colour in accordance with the instructions given in Chapter 3 of this Manual. Upon deciding the mixture strength at the throttle setting concerned, refer to the chart accompanying this text and note which of the three components listed (ie, main jet, jet needle and pilot air screws) need to be reset or replaced in order to weaken or richen the mixture strength over the throttle setting as desired. Note that the divisions shown in the chart indicate a certain amount of overlap between the various stages. Follow the instructions given in the following table when resetting or replacing the component concerned, taking care to err slightly on the side of a rich mixture, since a weak mixture will cause the engine to overheat with the subsequent risk of seizure.

| Component | Weaker mixture | Richer mixture |
|---|---|---|
| Main jet | Fit smaller No jet | Fit larger No jet |
| Jet needle | Move clip towards top of needle to lower needle | Move clip towards tip of needle to raise needle |
| Pilot air screw | Turn screw anti clockwise (out) | Turn screw clockwise (in) |

4   Finally, note that if the machine is found to perform badly when on ½ throttle setting, it is advisable to make any necessary adjustment to mixture strength by altering the settings of the pilot air screw and jet needle as opposed to changing the main jet.

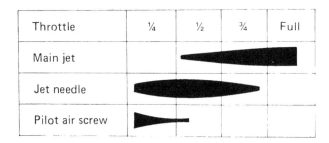

| Throttle | ¼ | ½ | ¾ | Full |
|---|---|---|---|---|
| Main jet | | | | |
| Jet needle | | | | |
| Pilot air screw | | | | |

**Fig. 2.2 Carburettor mixture strength chart**

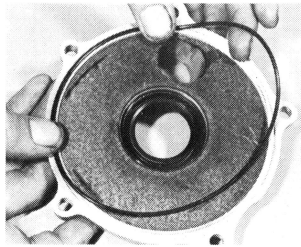

10.2a Check the condition of the disc valve fibre seat, the large O-ring, the seal ...

10.2b ... and the small O-ring

## 10 Disc valve induction system : examination and renovation

1   As mentioned in Section 1 of this Chapter, a disc valve, or rotary valve, arrangement is employed to give more precise control of the induction timing. This gives the advantages of more efficient combustion, with improved power output and fuel economy. It is most unusual for this unit to give any trouble, as it rarely shows any marked degree of wear during the normal service life of the machine. However, should any problem arise, the assembly may be removed after dismantling the clutch and primary drive assemblies. Refer to Chapter 1, Sections 11, 14, 36 and 39 for details.

2   Wear, and the need for renewal should be fairly obvious, such as worn splines in the disc centre and on the driving boss. The fibre valve seats may require renewal if badly worn or damaged by foreign matter becoming trapped between them and the disc itself. Bear in mind the remarks concerning disc timing, as this is critical to the efficient running of the engine.

## 11 Air filter element : removal, examination, cleaning and fitting

1   The air filter assembly consists of a large oil-impregnated polyurethane foam filter element which is housed in a frame-mounted contained located just to the rear of the cylinder barrel. This container is connected to the engine crankcase by means of a large-diameter rubber hose and attached to the frame by means of a mounting bracket with two screws and washers. The element itself must be removed from its container at 2000 mile (3000 km) intervals for the purposes of examination and cleaning.

2   To remove the element, move to the left-hand side of the machine and unscrew the single retaining screw from the base of the housing cover. With this screw removed, the cover may be lifted up and away from the locating tab, which passes through its upper edge, and thus clear of the container. Press in on the element retaining plate and detach each of its retaining hooks. Remove the plate and withdraw the element.

3   Carry out a close inspection of the element. If the foam of the element shows signs of having become hardened with age or is seen to be very badly clogged, then it must be renewed. To clean the element, detach the foam from its metal frame and immerse it in a non-flammable solvent, such as white spirit, whilst gently squeezing it to remove any oil and dust. After cleaning, squeeze out the foam by pressing it between the palms of both hands and then allow a short time for any solvent remaining in the form to evaporate. Do not wring out the foam as this will cause damage and thus lead to the need for early renewal.

4   Reimpregnate the foam with clean SAE 20W/40 oil and gently squeeze out any excess. Fit the metal frame into the foam so that the end of the frame is well inside the open end of the foam. Locate the element in the housing so that it fits correctly and secure its retaining plate in position with the two retaining hooks. Refit the cover to the end of the container and secure it in position with the single retaining bolt. Great care must be taken, when positioning both the element and the end cover, to ensure that no incoming air is allowed to bypass the element. If this is allowed to happen, it will allow any dirt or dust that is normally retained by the element to find its way into the carburettor and crankcase assemblies; it will also effectively weaken the fuel/air mixture.

5   Note that if the machine is being run in a particularly dusty or moist atmosphere, then it is advisable to increase the frequency of cleaning and reimpregnating the element. Never run the engine without the element fitted. This is because the carburettor is specially jetted to compensate for the addition of this component and the resulting weak mixture will cause overheating of the engine with the probable risk of severe engine damage.

11.2a Unscrew the single retaining bolt from the base of the air filter housing cover ...

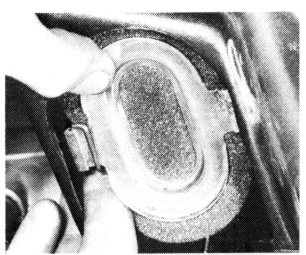

11.2b ... unclip the filter element retaining plate ...

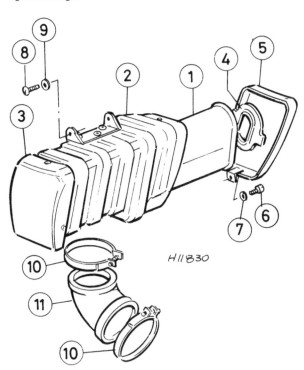

H11830

**Fig. 2.3 Air filter**

| | | | |
|---|---|---|---|
| 1 | Element | 7 | Spring washer |
| 2 | Element container | 8 | Screw – 2 off |
| 3 | Right-hand end cover | 9 | Washer – 2 off |
| 4 | Retaining plate | 10 | Hose clip – 2 off |
| 5 | Left-hand end cover | 11 | Rubber hose |
| 6 | Screw | | |

11.2c ... and withdraw the filter element

## 12 Engine lubrication system : general maintenance

1    As mentioned in Section 1 of this Chapter, the engine is pressure lubricated by an oil pump which is mounted to the rear of the left-hand crankcase half. it is recommended that adjustment to the control cable of this pump is checked every 2000 miles (3000 km), or on any occasion where the pump has been removed or where over or under lubrication are suspected. Full details of adjustment are contained in the following Section of this Chapter.

2    It is most important that an adequate level of oil is maintained in the frame-mounted oil tank. A sight glass is provided as a means of indicating that the oil level is low. Only the oil types listed in the Specifications Section of this Chapter should be used.

3    Whenever removing or carrying out adjustment to the pump, check the oil feed and delivery pipes for any signs of splitting or perishing and ensure that all connections and unions in the lubrication system are tight and free from leaks. Any fault found must be rectified immediately, as leakage will cause a loss of engine lubrication which will result in rapid wear of the engine components concerned or, at the worst, complete engine seizure.

## 13 Oil pump : removal and fitting

1    The oil pump can be expected to give long service, requiring no maintenance, but in the event of failure it must be renewed. No replacement parts are obtainable and the pump is, therefore effectively a sealed unit.

2    To gain access to the pump unit, first remove the gearchange lever by unscrewing its retaining bolts and then pulling it clear of the gearchange shaft end. Detach the rear section of the left-hand crankcase cover by removing its four securing screws. With the gearbox sprocket thus exposed, rotate the rear wheel until the split link in the final drive chain appears between the chain guard and the gearbox sprocket whilst on its upper run. Removal of the chain may now be achieved by removing the spring clip of the split link with a pair of flat-nose pliers and then withdrawing the link to allow the ends of the chain to separate. Lift the chain off the teeth of the sprocket and allow its ends to rest on a piece of clean rag or paper placed beneath the machine.

3    Remove the two screws that serve to retain the oil pump cover plate in position and manoeuvre the cover clear of the machine. Disconnect the pump control cable from the pump lever by pushing the end of the lever up so that tension is taken off the cable inner and then detaching the cable nipple from its nylon holder. With the rubber sealing cap detached from the cable adjuster, the cable may now be pulled through the adjuster so that it is clear of the pump.

4    It is now necessary to make provision for catching any oil that will issue from both the feed and delivery pipes once they are disconnected from the pump. To prevent complete draining of the oil tank, the feed pipe should be plugged as soon as it is disconnected; a clean screw or bolt of the appropriate thread diameter is ideal for this purpose. Slacken and remove both of the cross-headed retaining screws and lift the pump unit clear of its drive shaft end. Discard the pump base gasket and replace it with a new item.

5    To fit the replacement pump unit, clean both the pump and crankcase mating surfaces, place the new gasket onto the pump mating surface and align the central driven spigot of the pump with the slot in the end of its drive shaft before fitting the pump into its crankcase housing. With the pump correctly seated, fit and tighten to its two retaining screws. Note that each one of these screws must have a serviceable spring washer located beneath its head.

6    Unplug and reconnect both the oil feed and delivery pipes. Ensure that both pipes are a good push fit on their respective stubs and that the larger diameter pipe is correctly retained by its spring clip. Route both pipes correctly, taking care to ensure that they are neither twisted nor crimped between any component parts.

7    The oil pump must now be bled of air by following the instructions listed in the following Section of this Chapter.

8    Connect the pump control cable to the nylon holder in the pump lever end, remove the side cover from the carburettor housing and proceed to check the oil pump for correct adjustment in accordance with the following instructions.

9    Rotate the throttle twistgrip until the circular indicator mark on the base of the throttle slide comes into alignment with the upper edge of the carburettor mouth (see accompanying figure). With the throttle set in this position, check that the mark scribed on the pump lever boss is in exact alignment with the mark case in the pump body. If this is not the case, then the marks should be made to align by rotating the control cable adjuster, after having released the locknut. On completion of the adjustment procedure, retighten the locknut whilst holding the cable adjuster in position and slide the rubber sealing cap down the cable to cover the adjuster. It should be noted that any adjustment of the oil pump control cable may well effect the adjustment of the throtttle cable. It is, therefore, necessary to check the throttle cable for correct adjustment before proceeding further. Check that the gasket of the carburettor cover is in a satisfactory condition before refitting the cover and securing it in position with its three retaining screws. Note that any air allowed to enter through this joint will cause the fuel/air mixture strength to become weakened, thus adversely effecting the performance of the machine.

10   Refit the oil pump cover plate in position and fit and tighten its two retaining screws. Loop the final drive chain around the gearbox sprocket and reconnect its two ends with the split link. It is most important that the spring clip of this link is correctly fitted with the closed end facing the normal direction of chain travel. It is quite possible that the rear wheel will have to be moved forward in order to place enough slack on the chain to allow insertion of the link. In either case, the chain must be checked for correct tension and adjusted accordingly by referring to the instructions given in Chapter 5 of this Manual.

11   With the final drive chain refitted and correctly tensioned, place the rear section of the left-hand crankcase cover in position and secure it by fitting and tightening its four retaining screws. The gearchange lever can now be slid into position on the gearchange shaft. Ensure that the lever is positioned correctly in relation to the footrest and that its bolt hole is aligned with the channel in the shaft spline before inserting and tightening its retaining bolt. It was noticed on the model used for this Manual, that the lever retaining bolt had begun to work loose. It was, therefore, considered a good idea to apply a coating of thread locking compound to the threads of the bolt in order to prevent a recurrence of this problem. Finally, check that the oil tank is topped-up with oil of the specified type before starting the engine and allowing it to idle for a few minutes.

## 14 Oil pump : bleeding of air

1    If the oil pump has been removed, or the oil system drained, then it is most important to bleed any air from the system before the engine is started. Failure to do this will result in the engine seizing, with the resulting expense of a complete engine rebuild and the potential danger to the rider of the machine should the rear wheel lock unexpectedly whilst the machine is in motion.

2    Bleeding is effected by the cross-headed screw on the side of the pump body and waiting for a steady, air-free, stream of oil to emerge before retightening the screw.

12.2 A low level of engine oil will be seen through the sight glass

13.3 Remove the oil pump cover plate

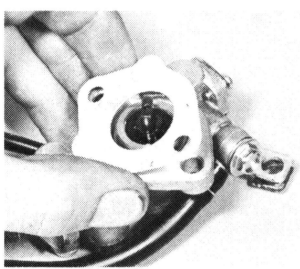

13.5 Always renew the oil pump base gasket

13.8 Reconnect the oil pump control cable

13.9a Rotate the oil pump control cable adjuster ...

13.9b ... until the marks on the pump lever and pump body align

14.2 Bleed the oil pump of air by loosening the bleed screw

## 15 Gearbox lubrication : general maintenance

1    General maintenance for the gearbox lubrication system consists solely of checking the level of oil within the gearbox casing at frequent intervals and changing the oil after every 2000 miles (3000 km). Carrying out these two service procedures will preclude any risk of the gearbox components becoming starved of oil or having to run in oil that has deteriorated to the point where it is ineffective in its prime functions as a lubricating medium.

2    Full details of checking the oil level and of carrying out an oil change can be found in Routine Maintenance at the front of this Manual.

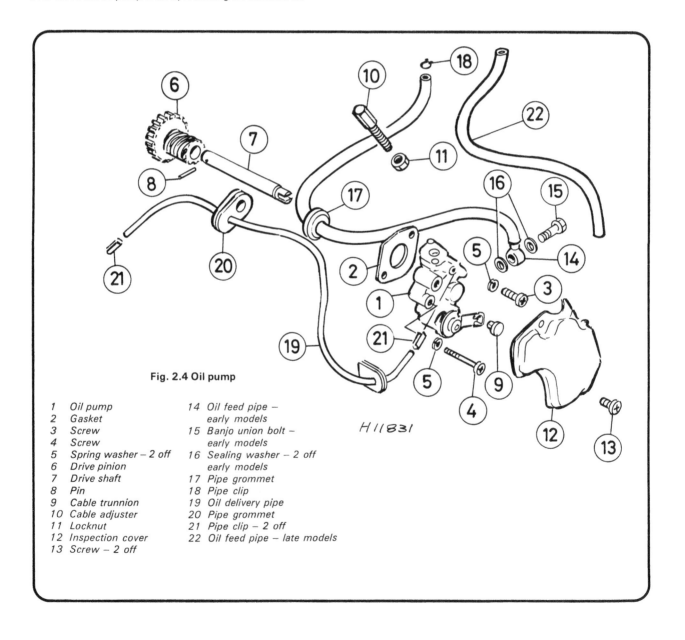

Fig. 2.4 Oil pump

| 1  | Oil pump |
| 2  | Gasket |
| 3  | Screw |
| 4  | Screw |
| 5  | Spring washer – 2 off |
| 6  | Drive pinion |
| 7  | Drive shaft |
| 8  | Pin |
| 9  | Cable trunnion |
| 10 | Cable adjuster |
| 11 | Locknut |
| 12 | Inspection cover |
| 13 | Screw – 2 off |

| 14 | Oil feed pipe – early models |
| 15 | Banjo union bolt – early models |
| 16 | Sealing washer – 2 off early models |
| 17 | Pipe grommet |
| 18 | Pipe clip |
| 19 | Oil delivery pipe |
| 20 | Pipe grommet |
| 21 | Pipe clip – 2 off |
| 22 | Oil feed pipe – late models |

## 16 Exhaust system : decarbonisation

1   After a considerable mileage has been covered, it will be necessary to decarbonise the complete exhaust system. The component most likely to require attention is the silencer baffle, which will tend to become choked if not kept clear. A two-stroke engine is very susceptible to this fault which is caused by the oily nature of the exhaust gases. As the sludge builds up back pressure will increase with a resulting fall off in performance.

2   It is not necessary to remove the exhaust system in order to gain access to the baffle. This is retained in the end of the silencer by a screw which may be removed by inserting a socket, box spanner or screwdriver through the hole in the underside of the silencer body, close to the end. With the screw removed, grip the end of the baffle with a pair of pliers or mole wrench and withdraw it from position. It will be found that, more often than not, both the baffle retaining screw and the baffle itself will be stuck fast in position due to the build up of carbon in the silencer assembly. In practice, the most satisfactory way found of freeing these components was to use a close-fitting socket on the head of the screw and a mole wrench, in conjunction with a tommy bar and hammer, on the end of the baffle. The baffle proved to be by far the most stubborn of the two components to free and was eventually pulled clear of the silencer by passing the tommy bar through the fitted mole wrench and striking the bar with the hammer, to shock the baffle from position.

3   If the build up of carbon and oil on the baffle is not too great, a wash with a fuel/paraffin mix will probably suffice to clean the component. Remember to take the usual precautions against fire when employing this cleaning method. When removing this component for the first time, it will be found that the rearmost end of the baffle is wrapped in an asbestos tape. This tape is not essential to the efficient operation of the silencer and should be cut away from the baffle and discarded.

4   If the build up of carbon and oil on the baffle is heavy, then more drastic action will be needed to clear away the accumulated deposits. It should be noted that a heavily contaminated baffle may well indicate the presence of similar contamination within the silencer casing and exhaust pipe assembly, in which case the complete exhaust system should be removed from the machine for inspection and cleaning as described in the following paragraphs of this Section. The most efficient method of removing heavy contamination from the baffle is to burn away the deposits by running the flame of a blowlamp or welding torch along the length of the baffle. On completion of this burning process, and with the baffle fully cooled, tap the baffle along its length with a piece of hard wood to dislodge any remaining deposits of carbon before giving it a final clean with a wire-bristled brush. Check that all the baffle holes are unobstructed.

5   To clean the silencer casing and exhaust pipe assembly, remove the system from the machine and suspend it from its silencer end. Block up the end of the exhaust pipe with a cork or wooden bung. If wood is used, allow an outside projection of three or four inches with which to grasp the bung for removal.

6   The method used to remove heavy carbon deposits from within the system is to dissolve them by using a chemical solution. Caustic soda dissolved in water is the solution most usually utilised as it is highly effective and relatively cheap. The mixture used is a ratio of 3 lbs caustic soda to over a gallon of fresh water. This is the strongest solution ever likely to be required. Obviously the weaker the mixture the longer the time required for the carbon to be dissolved. Note, whilst mixing the solution, that the caustic soda should be added to the water gradually, whilst stirring. Never pour water into a container of caustic soda powder or crystals; this will cause a violent reaction to take place which will result in great danger to one's person.

7   Bear in mind that it is very important to take great care when using caustic soda as it is a very dangerous chemical.

Always wear protective clothing, this must include proper eye protection. If the solution does come into contact with the eyes or skin it must be washed clear immediately with clean, fresh, running water. In the case of an eye becoming contaminated, seek expert medical advice immediately. Also, the solution must not be allowed to come into contact with aluminium alloy — especially at the above recommended strength — caustic soda reacts violently with aluminium and will cause severe damage to the component.

8   Commence the cleaning operating by pouring the solution into the system until it is quite full. Do not plug the open end of the system. The solution should now be left overnight for its dissolving action to take place. Note that the solution will continue to give off noxious fumes throughout its dissolving process; the system must therefore be placed in a well ventilated area. After the required time has passed, carefully pour out the solution and flush the system through with clean, fresh water. The cleaning operation is now complete.

9   Note the information contained in the following Section of this Chapter before refitting the exhaust system to the machine. When refitting the baffle in the silencer, ensure that the threaded hole in the baffle is aligned with the corresponding hole in the silencer before tapping the baffle into place. Check the condition of, and if necessary renew, the spring washer fitted beneath the head of the baffle retaining screw before fitting and tightening the screw. It is a good idea to prevent further seizure of this screw by lightly smearing its threads with graphite grease before insertion. Note that if this screw is not fitted or tightened properly and as a result of this, falls out, then the baffle will work loose, creating excessive exhaust noise accompanied by a marked fall off in performance.

10   Do not run the machine without the baffles in the silencer or modify the baffles in any way. Although the changed exhaust note may give the illusion of greater power, the chances are that the performance will fall off, accompanied by a noticeable lack of acceleration. The carburettor is jetted to take into account the fitting of silencers of a certain design and if this balance is disturbed the carburation will suffer accordingly.

## 17 Exhaust system : removal and fitting

1   The exhaust system can be removed from the machine as a complete assembly by first unscrewing the two bolts which serve to retain the exhaust pipe end to the cylinder barrel. Check that the spring washer located beneath each bolt head has not become flattened; if this is the case, the washers should be renewed before fitting takes place. Remove the locknut from the end of the swinging arm fork pivot shaft and pull the silencer bracket clear of the shaft end. The complete exhaust system may now be manoeuvred clear of the machine by lifting the end of the silencer and threading the exhaust pipe end between the engine and footrest and rear brake pedal assemblies. Under no circumstances should the exhaust system be allowed to hang from the cylinder barrel mounting whilst disconnected from the swinging arm shaft as this will impose an unacceptable strain on the threads of the two mounting bolts.

2   Fitting of the exhaust system is a direct reversal of the removal procedure, whilst noting the following points. Place a new sealing ring in the recess provided in the cylinder barrel before fitting the system into position. The spring washers fitted beneath the pipe to barrel retaining bolt heads and must be in good condition and the nut of the swinging arm fork pivot shaft should be tightened to a torque loading of 45 – 7.0 kgf m (32.5 – 50.5 lbf ft).

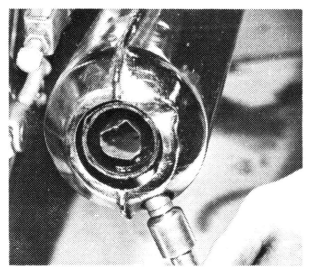

16.2a Remove the exhaust baffle retaining screw ...

16.2b ... and withdraw the baffle from the silencer

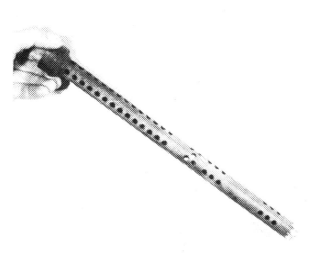

16.4 Ensure that none of the baffle holes are obstructed with carbon

17.2a Place a new sealing ring on the cylinder barrel recess ...

17.2b ... fit serviceable spring washers beneath the heads of the exhaust pipe retaining bolts ...

17.2c ... and tighten the swinging arm pivot nut to the correct torque loading

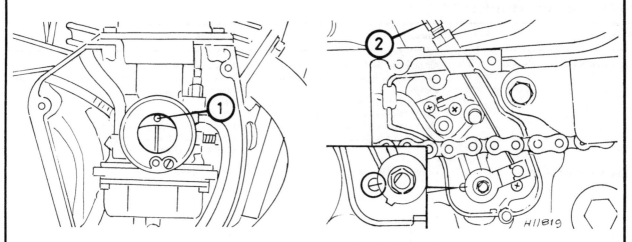

**Fig. 2.5 Oil pump adjustment synchronisation marks**

1   Carburettor punch mark

2   Cable adjuster

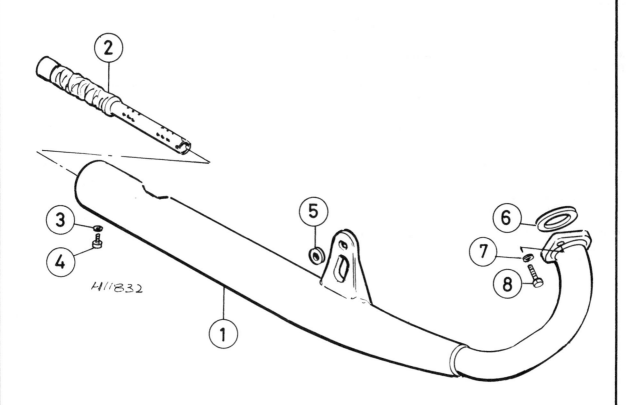

**Fig. 2.6 Exhaust system**

| | | | | | |
|---|---|---|---|---|---|
| 1 | Exhaust system | 4 | Screw | 7 | Spring washer – 2 off |
| 2 | Baffle | 5 | Washer | 8 | Bolt – 2 off |
| 3 | Spring washer | 6 | Sealing ring | | |

## 18  Fault diagnosis : fuel system

| Symptom | Cause | Remedy |
|---|---|---|
| Excessive fuel consumption | Air cleaner choked or restricted<br>Fuel leaking from carburettor. Float sticking<br>Badly worn or distorted carburettor<br>Jet needle setting too high<br>Main jet too large or loose<br>Carburettor flooding | Clean or renew.<br>Check all unions and gaskets. Float needle seat needs cleaning.<br>Replace.<br>Adjust as figure given in Specifications.<br>Fit correct jet or tighten if necessary.<br>Check float valve and replace if worn.<br>Check float height. |
| Idling speed too high | Throttle stop screw in too far<br>Carburettor top loose<br>Throttle cable sticking | Adjust screw.<br>Tighten.<br>Disconnect and lubricate or replace. |
| Engine dies after running for a short while | Blocked air hole in filler cap<br>Dirt or water in carburettor<br>Blocked fuel tap filter element | Clean.<br>Remove and clean out.<br>As above. |
| General lack of performance | Weak mixture: float needle stuck in seat<br>Air leak at carburettor point<br>Air leak at crankcase seal<br>Blocked air filter | Remove float chamber or float and clean.<br>Check joint to eliminate leakage.<br>As above.<br>Clean and relubricate element. |
| Engine sluggish. Does not respond to throttle | Back pressure in silencer<br><br>Fuel octane rating incorrect<br>Throttle cable sticking | Check baffles and clean if necessary.<br>Use correct grade (star rating) fuel<br>See above. |

## 19  Fault diagnosis : lubrication system

| Symptom | Cause | Remedy |
|---|---|---|
| White smoke from exhaust | Too much oil | Check oil pump setting and reduce if necessary. |
| Engine runs hot and gets sluggish when warm | Too little oil | Check oil pump setting and increase if necessary. |
| Engine runs unevenly, not particularly responsive to throttle openings | Intermittent oil supply | Bleed oil pump to displace air in feed pipes. |
| Engine drives up and seizes | Complete lubrication failure | Check for blockages in feed pipes, also whether oil pump drive has sheared. |

# Chapter 3 Ignition system

## Contents

## Specifications

| Ignition system | GP100 | GP125 |
|---|---|---|
| Type ...................................................................... | Coil and contract breaker | |

**Ignition timing**

Static ....................................................................... 20° (1.87 mm/0.074 in piston position) BTDC

Tolerance range ....................................................... ± 2° (1.52 - 2.26 mm/0.06 – 0.09 in piston position)

**Contact breaker**

Gap .......................................................................... 0.35 ± 0.05 mm (0.014 ± 0.002 in)

Dwell angle .............................................................. 170°

**Ignition coil:**

| | GP100 | GP125 |
|---|---|---|
| Primary winding resistance .................................. | 0.3 ohm | 1.5 ohm |
| Secondary winding resistance .............................. | 16 K ohm | 15 K ohm |

**Condensor**

Capacity:
Nippon Denso ....................................................... 0.18 ± 0.02 microfarad
Kokusan ................................................................ 0.25 ± 0.03 microfarad

**Spark plug**

Make ........................................................................ NGK or ND
Type ......................................................................... B8HS or W24FS
Gap .......................................................................... 0.6 – 0.8 mm (0.024 – 0.031 in)

**Flywheel generator**

| | GP100 | GP125 |
|---|---|---|
| Ignition source coil resistance ............................. | 0.1 ohm | 0.05 ohm |
| Charging coil resistance ...................................... | 0.2 ohm | 0.2 ohm |
| Lighting coil resistance ....................................... | 0.3 ohm | 0.2 ohm |
| Charging rate: | | |
| Lights on: | | |
| At 4000 rpm ......................................... | Above 0.6 amp | Above 0.7 amp |
| At 8000 rpm ......................................... | Below 2.8 amp | Below 1.5 amp |
| Lights off: | | |
| At 4000 rpm ......................................... | Above 0.7 amp | Above 0.8 amp |
| At 8000 rpm ......................................... | Below 2.8 amp | Below 2.5 amp |

## 1  General description

The conventional contact breaker ignition system fitted to the versions of the GP100 and GP125 models covered in this Manual, functions as follows. As the flywheel generator rotor operates, alternating current (ac) is generated in the ignition source coil mounted on the generator stator. Because the contact breaker is closed, the power runs to earth. When the contact breaker points open, the current is transferred to the primary windings of the ignition coil. In doing this, a high voltage is produced in the ignition secondary windings (by means of mutual induction) which is fed to the spark plug via the HT lead. As the energy flows to earth across the spark plug electrodes, a spark is produced and the combustible gases in the cylinder ignited. A capacitor (condenser) is fitted into the system to prevent arcing across the points; this helps reduce erosion due to burning.

## 2  Contact breaker points: adjustment

1  Access to the contact breaker assembly can be gained by removing the inspection cover from the forward section of the left-hand crankcase cover. This will reveal the flywheel generator rotor, which has four elongated holes in its outer face. The larger two of these holes are provided to permit inspection and adjustment of the contact breaker points.
2  Using the flat of a small screwdriver, open the contact breaker points against the pressure of the return spring so that the condition of the point contact faces may be checked. A piece of stiff card or crokus paper may be used to remove any light surface deposits, but if burnt or pitted, the complete contact breaker assembly must be removed to facilitate further examination of the point contact faces, and if necessary, renewal of the assembly. Refer to Section 4 of this Chapter for full details on removal, renovation and fitting of the assembly. If the points are found to be in sound condition, then proceed with adjustment as follows.
3  Rotate the crankshaft slowly until the contact breaker points are seen to be in their fully open position. Measure the gap between the points with a feeler gauge. If the gap is correct, a gauge of 0.35 mm (0.014 in) thickness will be a light sliding fit between the point faces. If this is not the case, slacken the single crosshead screw which serves to retain the fixed contact in position, just enough to allow movement of the contact. With the flat of a screwdriver placed in the indentation provided in the edge of the fixed contact plate, move the plate

in the appropriate direction until the gap is correct. Retighten the retaining screw and recheck the gap setting; it is not unknown for this setting to alter slightly upon retightening of the screw.
4  Prior to refitting the inspection cover, apply one or two drops of light machine oil to the cam lubricating wick whilst taking care not to allow excess oil to foul the point contact surfaces. Renew the cover sealing gasket if it is in any way damaged.

## 3  Ignition timing: checking and resetting

1  It cannot be overstressed that optimum performance of the engine depends on the accuracy with which the ignition timing is set. Even a small error can cause a marked reduction in performance and the possibility of engine damage as the result of overheating. Prior to checking the ignition timing the contact breaker gap should be checked as described in the following Section.
2  Static timing of the ignition can be carried out simply by aligning the timing mark scribed on the wall of the flywheel generator rotor with the corresponding mark on the crankcase and then checking that the contact breaker points are just on the point of separation. In order to gain access to the wall of the rotor, it will be necessary to remove the forward section of the left-hand crankcase cover.
3  Before commencing a check of the ignition timing, refer to the previous Section of this Chapter and check that the contact breaker points are both clean and correctly gapped. In order to provide an accurate indication as to when the contact breaker points begin to separate, it will be necessary to obtain certain items of electrical equipment. This equipment may take the form of a multimeter or ohmmeter, or a high wattage 6 volt bulb, complete with three lengths of electrical lead, which will be used in conjunction with the battery of the motorcycle.
4  When carrying out the above described method of static timing and the method of dynamic timing described at the end of this Section, note that the accuracy of both methods of timing depends very much on whether the flywheel generator rotor is set correctly on the crankshaft. Any amount of wear between the keyways in both the crankshaft and rotor bore and the Woodruff key will cause some amount of variation between the timing marks which will, in turn, lead to inaccurate timing. Inaccuracy in the timing mark position may also be a result of manufacturing error. The only means of overcoming this is to remove any movement between the two components and then to set the piston at a certain position within the cylinder bore

2.3a Measure the gap between the contact breaker points with a feeler gauge ...

2.3b ... and, if necessary, move the fixed contact to reset the gap before checking that the timing marks have remained in correct alignment. In order to accurately position the piston, it will be necessary to remove the spark plug and replace it with a dial gauge or a slide gauge either one of which is adapted to fit into the spark plug hole of the cylinder head.

5    Position the piston in the cylinder bore by first rotating the crankshaft until the piston is set in the top dead centre (TDC) position. Set the gauge at zero on that position and then rotate the crankshaft backwards (clockwise) until the piston has passed down the cylinder bore a distance of at least 4 mm (0.16 in). Reverse the direction of rotation of the crankshaft until the piston is exactly 1.87 mm (0.074 in) from TDC. The timing marks should now be in exact alignment. If this is not the case, then new marks will have to be made. Where the rotor has more than one timing mark highlight the mark that is in alignment with the crankcase mark. If none of the rotor marks are in alignment scribe a new mark on the rotor. All subsequent adjustment of the timing may be made using these marks.

6    Commence a check of the ignition timing by tracing the electrical lead from the fixed contact point (colour code, Black/Yellow) to a point where it can be disconnected. To set up a multimeter, set it to its resistance function, connect one of its probes to the lead end and the other probe to a good earthing point on the crankcase. Use a similar method to set up an ohmmeter. In each case, opening of the contact breaker points will be indicated by a deflection of the instrument needle from one reading of resistance to another.

7    When using the battery and bulb method, remove the battery from the machine and position it at a convenient point next to the left-hand side of the engine. Connect one end of a wire to the positive (+) terminal of the battery and one end of another wire to the negative (-) terminal of the battery. The negative lead may now be earthed to a point on the engine casing. Ensure the earth point on the casing is clean and that the wire is positively connected; a crocodile clip fastened to the wire is ideal. Take the free end of the positive lead and connect it to the bulb. The third length of wire may now be connected between the bulb and the electrical lead of the fixed contact point. As the final connection is made, the bulb should light with the points closed. Opening of the contact breaker points will be indicated by a dimming of the bulb. The reason for recommending the use of a high voltage bulb is that this dimming will be more obvious to the eye.

8    Ignition timing is correct when the contact breaker points are seen to open just as the timing marks come into alignment. If this is not the case, then the points will have to be adjusted by moving the stator plate clockwise or anti-clockwise, depending on whether the opening point needs to be advanced or retarded. This is accomplished by removing the flywheel generator rotor in order to gain full access to the three stator plate retaining screws, each one of which passes through an elongated slot cut in the plate.

9    Removal of the flywheel rotor may be achieved by carrying out the following procedure. Prevent the crankshaft from rotating by selecting top gear and then applying the rear brake. Remove the rotor retaining nut, spring washer and plain washer. The rotor is a tapered fit on the crankshaft end and is located by a Woodruff key; it therefore requires pulling from position. The rotor boss is threaded internally to take the special Suzuki service tool No 09930 – 30102 with an attachment, No 09930 – 30161. If this tool cannot be acquired, then it is possible to remove the rotor by careful use of a two-legged puller.

10  If it is decided to make use of a two-legged puller, great care must be taken to ensure that both the threaded end of the crankshaft and the rotor itself remain free from damage. To protect the crankshaft end, refit the rotor retaining nut and screw it on until its outer face lies flush with the end of the crankshaft. Assemble the puller, checking that its feet are fitted through the two larger slots in the rotor face and are resting securely on the strengthened hub. Remember that the closer the feet of the puller are to the centre of the rotor then the less chance there is of the rotor becoming distorted. Gradually tighten the centre bolt of the puller to apply pressure to the rotor. Do not overtighten this bolt but apply reasonable pressure and then strike the end of the bolt with a hammer to break the taper joint. If this fails at first, tighten the centre bolt to apply a little more pressure and then try shocking the rotor free again.

11  It will be appreciated that resetting of the timing is a repetitive process of loosening the three stator plate retaining screws, rotating the stator plate a small amount in the required direction, retightening the screws, pushing the rotor back onto the crankshaft taper and then rechecking the timing. With the ignition timing correctly set, check tighten the stator plate retaining screws and then refit the rotor by first cleaning and degreasing both the taper of the crankshaft and the bore of the rotor where the two components come into contact. Check that the Woodruff key is correctly fitted in the crankshaft keyway and then push the rotor onto the crankshaft. Gently tap the centre of the rotor with a soft-faced hammer to seat it on the crankshaft taper and then fit the plain washer followed by the spring washer. Clean and degrease the threads of the rotor retaining nut and of the crankshaft end. Apply a thread locking compound to these threads and fit and tighten the nut, finger-tight. Lock the crankshaft in position by employing the method used for rotor removal and tighten the rotor retaining nut to a torque loading of 3.0 – 4.0 kgf m (21.5 – 29.0 lbf ft).

12  An alternative method of checking the ignition timing can be adopted, whilst the engine is running, using a stroboscopic lamp. This will entail gaining access to the wall of the rotor by removing the forward section of left-hand crankcase cover. When the light from the lamp is aimed at the timing marks on the crankcase and rotor wall, it has the effect of 'freezing' the moving mark on the rotor in one position and thus the accuracy of the timing can be seen.

13  Prepare the timing marks by degreasing them and then coating each one with a trace of white paint. This is not absolutely necessary but will make the position of each mark far easier to observe if the light from the 'stroke' is weak or if the timing operation is carried out in bright conditions. If the rotor has more than one timing mark the timing must first be checked statically to determine the correct mark, see paragraph 5. Once the correct mark has been established, and highlighted with a trace of white paint, all successive timing checks can be made dynamically using a stroboscope.

14  Two basic types of stroboscopic lamp are available, namely the neon and xenon tube types. Of the two, the neon type is much cheaper and will usually suffice if used in a shaded position, its light output being rather limited. The brighter but more expensive xenon types are preferable, if funds permit, because they produce a much clearer image.

15  Connect the 'strobe' to the HT lead, following the maker's instructions. If an external 6 volt power source is required, use the battery from the machine but make sure that when the leads from the machine's electrical system are disconnected from the terminals of the battery, they have their ends properly insulated with tape in order to prevent them from shorting on the cycle components.

16  Start the engine and aim the 'strobe' at the timing marks. The ignition timing is correct when the mark on the crankcase is in exact alignment with the mark on the rotor wall. To adjust the ignition timing, follow the instructions given in paragraphs 8, 9, 10 and 11 of this Section.

17  Finally, on completion of either one of the above mentioned timing procedures, refit the disturbed rotor cover and remove all test equipment from the machine.

## 4  Contact breaker assembly: removal, renovation and fitting

1    If the contact breaker points are found to be burned, pitted or badly worn, they should be removed for dressing. If, however, it is necessary to remove a substantial amount of material before the faces can be restored, new contacts should be fitted.

2   The contact breaker assembly forms part of the flywheel generator stator plate assembly. To gain full access to this assembly, it is necessary to remove the crankcase covers surrounding the flywheel generator and then to remove the rotor. Full instructions for the removal and fitting of these components are given in the relevant Sections of Chapter 1.

3   The contact breaker assembly is secured to the stator plate by means of a single crosshead screw with a spring washer beneath its head. Before the assembly can be removed, it is first necessary to detach both the condenser and coil leads from the spring blade of the moving point. They are retained by a single crosshead screw with spring washer, which should be relocated immediately after the leads have been detached in order to avoid loss.

4   Using the flat of a small electrical screwdriver, prise off the circlip which retains the moving contact assembly to its pivot pin. Remove the plain washer, followed by the moving contact complete with insulating washers. Make a note of the order in which components are removed, as they are easily assembled incorrectly. Release the fixed contact by unscrewing the single retaining screw which passes through its backplate. Do not, under any circumstances, loosen the three stator mounting screws, otherwise the ignition timing will be lost.

5   The points surfaces may be dressed by rubbing them on an oilstone or fine emery paper, keeping the points square to the abrasive surface. If possible, finish off by using crokus paper to give a polished surface, which is less prone to subsequent pitting. Make sure all traces of abrasive are removed before reassembly.

6   Reassemble the contact breaker assembly by reversing the dismantling sequence, taking care that the insulating washers are replaced correctly. If this precaution is not observed, it is easy to inadvertently earth the assembly rendering it inoperative. The pivot pin should be greased sparingly, and a few drops of oil applied to the cam lubricating wick.

7   If the contact breaker is being renewed due to excessive burning of the contacts, this is likely to have been caused by a faulty condenser. Refer to Section 7 of this Chapter if this is suspected.

8   On completion of reassembling and fitting the renovated or renewed contact breaker assembly and after having refitted the generator rotor, adjust the contact breaker gap and check the ignition timing, as detailed in the relevant Sections of this Chapter.

## 5   Flywheel generator: checking the output

1   The flywheel generator is instrumental in creating the power in the ignition system, and any failure or malfunction will affect the operation of the ignition system. If the machine will not start and there is no evidence of a spark at the spark plug, a check should first be made to ensure there is no fault in either the spark plug itself, the suppressor cap, the HT coil and lead assembly, the contact breaker assembly or the condenser.

2   If the above listed checks prove satisfactory, then the output from the generator itself should be suspected. Refer to Section 4 of Chapter 6 for the output checking procedure by basic test methods, but before doing this, a thorough check of the circuit wiring should be made to ensure the wiring connections are not badly corroded or contaminated by dirt or moisture. Check the wiring itself for signs of chafing against the frame or engine components or any indication of a break in the wiring.

## 6   Ignition coil – location and testing

1   The ignition coil fitted to the machines covered in this Manual is a sealed unit which is designed to give long and trouble-free service without need for attention. The coil is located beneath the fuel tank and is mounted beneath the frame top tube, just to the rear of the steering head assembly. It follows therefore that the fuel tank must be removed in order to gain access to the coil. Refer to Section 2 of Chapter 2 for the tank removal and refitting procedures.

2   Bear in mind, before removing the fuel tank in order to gain access to the ignition coil, that a defective condenser in the contact breaker circuit can give the illusion of a defective coil and for this reason it is advisable to investigate the condition of the condenser before condemning the ignition coil. Refer to Section 7 of this Chapter for the appropriate details.

3   If, however, a weak spark and difficult starting causes the performance of the ignition coil to be suspect, then it should be tested as follows. Trace the thin Black/yellow electrical lead from the coil to its bullet connector and pull apart the connector. Disconnect the suppressor cap from the spark plug. Set a multimeter on its resistance function (K ohm scale). Connect one probe of the meter to earth and the other to the suppressor cap connection. Approximately the specified resistance (see Specifications) should be shown on the scale of the meter if the secondary windings of the coil are in good order.

4   To check the condition of the primary windings of the coil, reset the meter to its ohms scale and with one of its probes remaining on earth, connect the other to the thin Black/yellow lead. Approximately the specified resistance (see Specifications) should be indicated if the primary windings are in good order.

5   If the coil has failed, it is likely to have either an open or short circuit in its primary or secondary windings. This type of fault will be immediately obvious and will require the renewal of the coil. Where the fault is less clear cut, it is advisable to have the suspect coil tested on a spark gap tester by an official Suzuki service agent.

## 7   Condenser: testing, removal and fitting

1   A condenser is included in the contact breaker circuitry to prevent arcing across the contact breaker points as they separate. The condenser is connected in parallel with the points and if a fault develops, ignition failure is liable to occur.

2   If the engine proves difficult to start, or misfiring occurs, it is possible that the condenser is at fault. To check, separate the contact breaker points by hand when the ignition is switched on. If a spark occurs across the points and they have a blackened and burnt appearance, the condenser can be regarded as unserviceable.

3   It is not possible to check the condenser without the appropriate equipment. In view of the low cost involved, it is preferable to fit a new replacement and observe the effect on engine performance. The condenser may be removed and the new item fitted as follows.

4   The condenser forms part of the flywheel generator stator plate assembly. To gain full access to this assembly, it is necessary to remove the crankcase covers surrounding the flywheel generator and then to remove the rotor. Full instructions for the removal and fitting of these components are given in the relevant Sections of Chapter 1.

5   With the stator plate thus exposed, commence removal of the condenser by carefully easing back the lead retaining clip, which is situated on top of the component, until it is clear of the leads. It will now be necessary to place the heated top of a soldering iron on the union of the leads to the condenser in order to melt the soldered joint and thus free the leads. Detach the condenser from the stator plate by removing the single crosshead retaining screw with its spring washer.

6   The procedure adopted for fitting the new condenser should be an exact reversal of that used for removal of the unserviceable component. Take great care to ensure that the soldered joint is properly made and that each lead is properly routed through the retaining clip.

4.2 The contact breaker assembly and condenser are both mounted on the flywheel generator stator plate

6.1 The ignition coil is mounted beneath the frame top tube

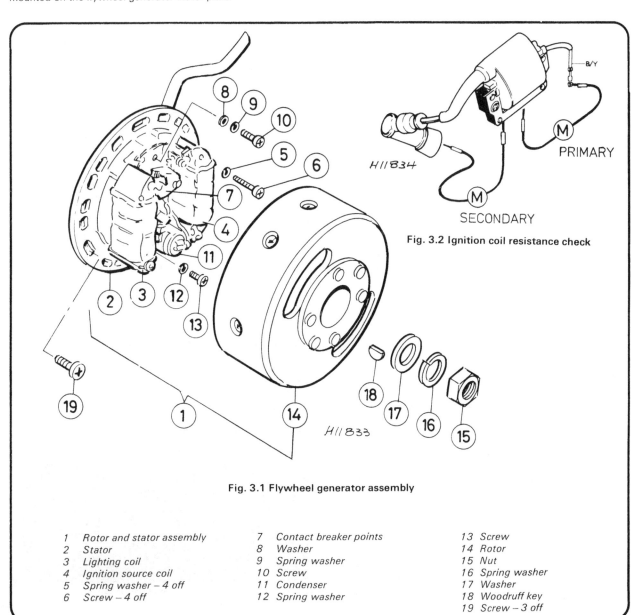

Fig. 3.2 Ignition coil resistance check

Fig. 3.1 Flywheel generator assembly

| | | |
|---|---|---|
| 1 Rotor and stator assembly | 7 Contact breaker points | 13 Screw |
| 2 Stator | 8 Washer | 14 Rotor |
| 3 Lighting coil | 9 Spring washer | 15 Nut |
| 4 Ignition source coil | 10 Screw | 16 Spring washer |
| 5 Spring washer – 4 off | 11 Condenser | 17 Washer |
| 6 Screw – 4 off | 12 Spring washer | 18 Woodruff key |
| | | 19 Screw – 3 off |

### 8  Spark plug: cleaning and resetting the gap

1   Suzuki fit an NGK B8HS (or ND W24FS) spark plug as standard equipment to all the machines covered in this Manual. The recommended gap between the plug electrodes is 0.6 – 0.8 mm (0.024 – 0.031 in). The plug should be cleaned and the gap checked and reset at the service interval recommended in the Routine Maintenance Chapter at the beginning of this Manual. In addition, in the event of a roadside breakdown where the engine has mysteriously 'died' the spark plug should be the first item checked.
2   The plug should be cleaned thoroughly by using one of the following methods. The most efficient method of cleaning the electrodes is by using a bead blasting machine. It is quite possible that a local garage or motorcycle dealer has one of these machines installed on the premises and will be willing to clean any plugs for a nominal fee. Remember, before fitting a plug cleaned by this method, to ensure that there is none of the blasting medium left impacted between the porcelain insulator and the plug body. An alternative method of cleaning the plug electrodes is to use a small brass-wire brush. Most motorcycle dealers sell such brushes which are designed specifically for this purpose. Any stubborn deposits of hard carbon may be removed by judicious scraping with a pocket knife. Take great care not to chip the porcelain insulator round the centre electrode whilst doing this. Ensure that the electrode faces are clean by passing a small fine file between them, alternatively, use emery paper but make sure that all traces of the abrasive material are removed from the plug on completion of cleaning.
3   To reset the gap between the plug electrodes, bend the outer electrode away from or closer to the central electrode and check that a feeler gauge of the correct size can be inserted between the electrodes. The gauge should be a light sliding fit. Never bend the central electrode or the insulator will crack, causing engine damage if the particles fall in whilst the engine is running.
4   With some experience, the condition of the sparking plug electrodes and insulator can be used as a reliable guide to engine operating conditions. See the accompanying colour photographs.
5   Always carry a spare spark plug of the correct type. The plug in a two-stroke engine leads a particularly hard life and is liable to fail more readily than when fitted to a four-stroke.

6   Beware of overtightening the spark plug, otherwise there is risk of stripping the threads from the aluminium alloy cylinder head. The plug should be sufficiently tight to seat firmly on its sealing washer, and no more. Use a spanner which is a good fit to prevent the spanner from slipping and breaking the insulator.
7   If the threads in the cylinder head strip as a result of overtightening the spark plug, it is possible to reclaim the head by use of a Helicoil thread insert. This is a cheap and convenient method of replacing the threads; most motorcycle dealers operate a service of this nature at an economic price.
8   Before fitting the spark plug in the cylinder head, coat its threads sparingly with a graphited grease. This will prevent the plug from becoming seized in the head and therefore aid future removal.
9   When reconnecting the suppressor cap to the plug, make sure that the cap is a good, firm fit and is in good condition; renew its rubber seals if they are in any way damaged or perished. The cap contains the suppressor that eliminates both radio and TV interference.

### 9  High tension lead: examination

1   Erratic running faults and problems with the engine suddenly cutting out in wet weather can often be attributed to leakage from the high tension lead and spark plug cap. If this fault is present, it will often be possible to see tiny sparks around the lead and cap at night. One cause of this problem is the accumulation of mud and road grime around the lead, and the first thing to check is that the lead and cap are clean. It is often possible to cure the problem by cleaning the components and sealing them with an aerosol ignition sealer, which will leave an insulating coating on both components.
2   Water dispersant sprays are also highly recommended where the system has become swamped with water. Both these products are easily obtainable at most garages and accessory shops. Occasionally, the suppressor cap or the lead itself may break down internally. If this is suspected, the components should be renewed.
3   Where the HT lead is permanently attached to the ignition coil, it is recommended that the renewal of the HT lead is entrusted to an auto-electrician who will have the expertise to solder on a new lead without damaging the coil windings.

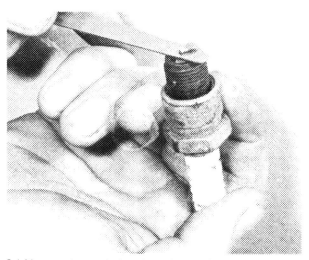

8.1 Measure the spark plug electrode gap with a feeler gauge

8.7 Ensure that the spark plug suppressor cap is in good condition

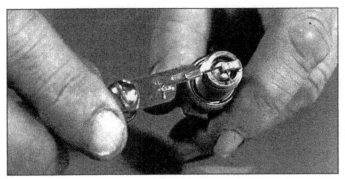

**Electrode gap check** - use a wire type gauge for best results

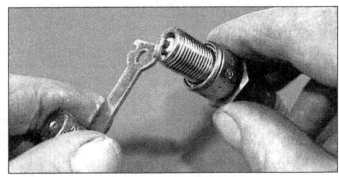

**Electrode gap adjustment** - bend the side electrode using the correct tool

**Normal condition** - A brown, tan or grey firing end indicates that the engine is in good condition and that the plug type is correct

**Ash deposits** - Light brown deposits encrusted on the electrodes and insulator, leading to misfire and hesitation. Caused by excessive amounts of oil in the combustion chamber or poor quality fuel/oil

**Carbon fouling** - Dry, black sooty deposits leading to misfire and weak spark. Caused by an over-rich fuel/air mixture, faulty choke operation or blocked air filter

**Oil fouling** - Wet oily deposits leading to misfire and weak spark. Caused by oil leakage past piston rings or valve guides (4-stroke engine), or excess lubricant (2-stroke engine)

**Overheating** - A blistered white insulator and glazed electrodes. Caused by ignition system fault, incorrect fuel, or cooling system fault

**Worn plug** - Worn electrodes will cause poor starting in damp or cold weather and will also waste fuel

## 10 Fault diagnosis: ignition system

| Symptom | Cause | Remedy |
|---|---|---|
| Engine will not start | No spark at plug | Faulty ignition switch. Check whether current is reaching ignition coil. |
| | Weak spark at plug | Dirty contact breaker points require cleaning. Contact breaker gap has closed up. Reset. |
| Engine starts, but runs erratically | Break or short in LT circuit | Locate and rectify. If no improvement check whether points are arcing. If so renew condenser. |
| | Ignition over-advanced | Check ignition timing and if necessary, reset. |
| | Plug lead insulation breaking down | Check for breaks in outer covering, especially near frame. |
| Engine difficult to start and runs sluggishly. Overheats | Ignition timing retarded | Check ignition timing and advance to correct setting. |

# Chapter 4 Frame and forks

## Contents

## Specifications

### Frame

| | |
|---|---|
| Type | Welded tubular steel, semi cradle, engine used as stressed member |

### Front forks

| | |
|---|---|
| Type | Oil damped, telescopic |
| Travel | 110 mm (4.32 in) |
| Spring free lenth | 496 mm (19.53 in) |
| Service limit | 481 mm (18.94 in) |
| Oil capacity per leg | 90 cc (3.17 Imp fl oz) |
| Oil type | SAE 10W/20 |

### Rear suspension

| | |
|---|---|
| Type | Swinging arm fork, controlled by two suspension units having internal oil-filled dampers |
| Travel | 85 mm (3.35 in) |
| Fork pivot shaft maximum runout | 0.6 mm (0.02 in) |

### Torque wrench settings

| | kgf m | lbf ft |
|---|---|---|
| Front fork cap bolt | 3.5 - 5.5 | 25.5 - 40.0 |
| Lower yoke pinch bolts | 2.5 - 3.5 | 18.0 - 25.5 |
| Steering stem top bolt | 3.5 - 5.5 | 25.5 - 40.0 |
| Handlebar clamp bolts | 1.2 - 2.0 | 8.5 - 14.5 |
| Swinging arm fork pivot shaft nut | 4.5 - 7.0 | 32.5 - 21.5 |
| Rear suspension unit securing nuts | 2.0 - 3.0 | 14.5 - 21.5 |
| Centre stand pivot bolt nut | 3.0 - 3.7 | 21.5 - 27.0 |

## 1 General description

The type of frame utilised on the models covered in this Manual is of a conventional welded tubular steel construction, the engine unit being bolted between the front downtube and the main frame structure to form a structural part of the frame.

The front forks are of the conventional telescopic type, having internal oil-filled dampers. The fork springs are contained within the fork stanchions and each fork leg can be detached from the machine as a complete unit, without dismantling the steering head assembly.

Rear suspension is of the swinging arm type, using oil filled suspension units to provide the necessary damping action. The units are adjustable so that the spring ratings can be effectively changed within certain limits to match the load carried.

## 2  Front fork legs: removal and fitting

1    Place the machine securely on its centre stand, leaving plenty of working area at the front and sides. Arrange wooden blocks beneath the crankcase so that the front wheel is raised clear of the ground. Remove the front wheel as instructed in Section 3 of the following Chapter.

2    On models fitted with a disc brake, unscrew and remove the two bolts which retain the brake caliper support bracket to the fork leg. Swing the caliper unit back from the fork leg and suspend it from a point on the frame by means of a length of wire or string. Ensure that the hardwood wedge placed between the brake pads during wheel removal remains in position. Note that the hydraulic hose need not be disconnected from the caliper unit.

3    Where necessary, disconnect the speedometer cable from the speedometer gearbox and thread it through its guide on the front mudguard. Secure the cable to a point on the frame clear of the fork assembly.

4    Detach the front mudguard from each of the fork legs and remove it from the machine. The mudguard is secured by four bolts, each with a spring washer, two of which thread into each fork leg.

5    It is now necessary to detach the handlebars from their mounting points on the upper fork yoke in order to allow removal of the fork cap bolts. The handlebars are retained in position by two sets of U-clamps, each set being secured to the upper yoke by two bolts with spring washers. Before unscrewing these bolts to allow removal of the handlebars, cover the forward section of the fuel tank with an old blanket, or similar, to protect its paint finish from damage. Detach the handlebars from the yoke and place them on the padded area of tank. Where a hydraulically operated disc brake is fitted, take great care to keep the brake reservoir upright to avoid any spillage of hydraulic fluid onto the tank surface or frame components. Hydraulic fluid will act as an efficient paint remover and will also cause damage to any plastic components.

6    Proceed to remove each fork leg by first removing its cap bolt and then slackening the clamp bolt which serves to retain the fork leg in the lower yoke. The leg can now be eased downwards out of position. If the clamp of the lower yoke proves to be excessively tight, then it may be gently sprung by using the flat of a large screwdriver. This must be done with great care in order to prevent breakage of the clamp which will necessitate replacement of the complete yoke.

7    Once removed, each fork leg can be dismantled for inspection and renovation as described in Section 3 of this Chapter.

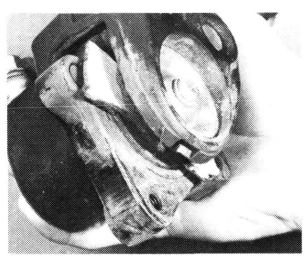

2.2 Remove the brake caliper unit from the fork lower leg

2.4 Detach the front mudguard from the fork leg ...

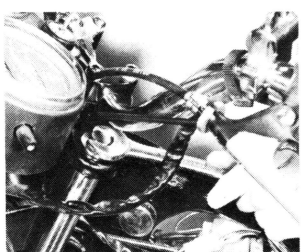

2.6a ... unscrew the fork leg cap bolt ...

2.6b ... slacken the lower yoke clamp bolt ...

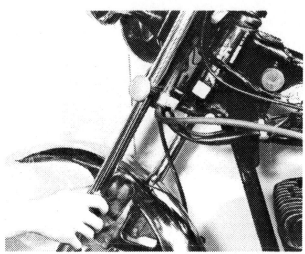

2.6c ... and pull the fork leg down clear of the steering head

8   Fitting of each fork leg is a straightforward reversal of the removal sequence, whilst noting the following points. Push the leg stanchion hard up against the upper yoke before refitting the cap bolt and nipping it tight to hold the leg in position. With both forks and their related components in position, fully tighten each component securing nut or bolt, starting with the wheel spindle retaining nut and working upwards. Take note of the torque settings given in the Specifications Section of this Chapter and of Chapter 5.

9   When refitting the handlebars, note that they must be fitted so that the punch mark on the handlebar is directly in line with the rear mating faces of the handlebar clamps. Ensure that the clamp retaining bolts are refitted with serviceable spring wash-ers beneath their heads and are tightened evenly so that the gap between the forward mating surfaces of the clamps is equal to that between the rear mating surfaces.

### 3   Front fork legs: dismantling, examination, renovation and reassembly

1   It is strongly advised that each fork leg is dismantled separately whilst using an identical procedure. If this approach is adopted, it will mean that there is less chance of component parts being unwittingly exchanged between fork legs.

2   Commence by draining each fork leg of damping oil. This can be achieved by clamping the fork leg stanchion between the jaws of a vice, whilst keeping the leg in a vertical plane, and unscrewing the fork spring retaining plug. The plug is under spring pressure, so care must be taken when unscrewing the plug the last few threads. With this plug removed, the fork spring(s) may be withdrawn from the stanchion, the leg removed from the vice and inverted over a suitable receptacle, and the oil allowed to drain. Move the lower leg of the fork up and down the stanchion several times to eject any oil remaining within the leg before placing the leg on a clean work surface ready for further dismantling work to be carried out. Note that whenever any part of a fork leg is clamped between the jaws of a vice, the contact surface of the jaws must be protected to avoid their knurled surfaces causing damage to either the polished surface of the stanchion or the soft alloy of the lower leg. Thin wooden blocks or soft aluminium alloy pieces are ideal for providing the necessary protection, whereas the use of rag is not advised as components are more liable to dip from position.

3   The next step in the dismantling procedure is to free the damper rod assembly by removing the Allen-headed bolt which fits in a recess in the base of the fork leg. In some cases the screw can be removed with ease, but it is likely that the damper rod, into which the bolt screws, will rotate. In order to hold the rod in position whilst the bolt was removed, it was found necessary to manufacture a special tool from a length of square-section steel bar. Alternatively, Suzuki special tools Nos 09940 – 34520 and 09940 – 34561 may be obtained for this purpose.

4   To manufacture the tool, obtain a length of steel bar which is approximately 0.5 inch thick, square in section and approximately 20 inches long. Cut one end of this bar to a point, as shown in the figure accompanying this text. It is now necessary to harden this point so that when it is pushed into position against the circular recess in the end of the damper rod, its edges will bite into the softer metal of the recess walls thus providing enough grip to prevent the rod from rotating. The procedure to bring the point to the required degree of hardness is as follows.

5   Using the flame of a blowlamp or welding torch, heat the bar to a cherry red for about half its length from the pointed end. Directly the bar turns red with the heat, quench its end in a large container full of water. Only 1-2 inches of the bar need be submerged beneath the surface of the water for this initial cooling procedure. Once the end of the bar has cooled, quickly remove it from the water and polish the edges of its pointed end with emery cloth. The heat remaining in the uncooled section of bar will soon travel, by conduction, to the tip of the point. As this happens, the spectrum of tempering colours will be seen to progress up the edges of the point. Upon seeing the straw colour appear on the point edges, quench the complete tool in the water until it is completely cooled. The bar should now have a hardened point with the metal gradually decreasing in hardness towards the mid point of its length.

6   When carrying out the procedure described above, take great care to observe the following safety precautions. Always be aware of the dangers which come from naked flames and heated metal; have a means of extinguishing fire nearby and wear both eye and body protection. Thrusting red hot metal into cold water will produce a very violent reaction between the two; be prepared for this and guard against the possibility of scalding water being thrown from the container.

7   Once the bar has cooled, decide upon a means of prevent-ing it from turning whilst the damper rod retaining bolt is being unscrewed. If a vice is available, then the problem is easily solved as all that needs to be done is to clamp the end of the bar between the jaws of the vice. Otherwise, clamp a self-grip wrench to the end of the bar or drill a hole through which to pass a tommy bar.

8   Locate the point of the special tool in the recess of the damper rod by passing it down inside the stanchion and giving its end a sharp tap. Unscrew the retaining bolt and withdraw the tool, followed by the damper rod assembly. In practice, it was found that several attempts were required in order to get the tool to grip the damper rod firmly but with patience and a great deal of application, this was eventually achieved.

9   Using the flat of a screwdriver, carefully prise the dust excluder from its location on the lower leg and slide it up and off the fork stanchion. The stanchion can now be pulled out of the lower leg.

10  The oil seal fitted within the top of the lower leg should be removed only if it is to be renewed. This is because damage will almost certainly be inflicted upon the seal as it is prised from position. The spring clip which retains the seal in position may be displaced by inserting the flat of a small screwdriver into one of the clip indentations provided. With the clip thus removed, the seal may be lowered out of position by placing the flat of a screwdriver beneath its lower edge. Take great care when removing both of these items not to damage the alloy edge of the fork leg with the screwdriver.

11  The type of fork legs fitted to the machines covered in this Manual do not contain bushes. The lower legs slide directly against the outer hard chrome surface of the fork stanchions. If wear occurs, indicated by slackness, the fork leg complete will

have to be renewed, possibly also the fork stanchion. Wear of the fork stanchion is indicated by scuffing and penetration of the hard chrome surface.

12 Check the outer surface of the stanchion for scratches or roughness, it is only too easy to damage the oil seal during the re-assembly if these high spots are not eased down. The stanchions are unlikely to bend unless the machine is damaged in an accident. Any significant bend will be detected by eye, but if there is any doubt about straightness, roll down the stanchion tubes on a flat surface such as a sheet of plate glass. If the stanchions are bent, they must be renewed. Unless specialised repair equipment is available it is rarely practicable to effect a satisfactory repair to a damaged stanchion.

13 After an extended period of service, the fork springs may take a permanent set. If the spring lengths are suspect, then they should be measured and the readings obtained compared with the service limits given in the Specifications Section of this Chapter. It is always advisable to fit new fork springs where the length of the original items has decreased beyond the service limit given. Always renew the springs as a set, never separately. Where there are two springs fitted within each fork leg, take note of the condition of the seating ring which separates the two springs. If this ring shows signs of excessive wear or damage, then it must be renewed.

14 The piston ring fitted to the damper rod may wear if oil changes at the specified intervals are neglected. If damping has become weakened and does not improve as a result of an oil change, the piston ring should be renewed. Check also that the oilways in the damper rod have not became obstructed. Suzuki provide no information as to the amount of set allowed on the damper rod spring before renewal is necessary. If in doubt as to the condition of this spring, ask the advice of a Suzuki service agent or compare the spring against a new item.

15 Closely examine the dust excluder for splits or signs of deterioration. If found to be defective, it must be renewed as any ingress of dirt will rapidly accelerate wear of the oil seal and fork stanchion. It is advisable to renew any gasket washers fitted beneath bolt heads as a matter of course. The same applies to the O-rings fitted to the cap bolts.

16 Reassembly of each fork leg is essentially a reversal of the dismantling procedure, whilst noting the following points. It is essential that all fork components are thoroughly washed in solvent and wiped clean with a lint-free cloth before assembly takes place. Any trace of dirt inside the fork leg assembly will quickly destroy the oil seal or score the stanchion to outer fork leg bearing surfaces.

17 With the damper rod inserted into the fork stanchion and both components secured in the lower fork leg by means of the Allen-headed retaining bolt with its sealing washer, commence

fitting of the new oil seal, where required. Before fitting the new seal, it should be coated with the recommended fork oil on its inner and outer surfaces. This serves to make the fitting of the seal into the lower fork leg easier and also reduces the risk of damage to the seal when it is passed over the stanchion. Take great care when fitting the seal over the stanchion.

18 Suzuki recommend the use of a service tool (No 09940-50110) with which to drive the seal into the fork leg recess. With the seal partially located in the leg recess the tool, which takes the form of a short length of metal tube approximately 3 in long, with an inner diameter just greater than the outer diameter of the stanchion and an outer diameter just less than that of the outer diameter of the oil seal, may be passed over the stanchion and used to tap the seal home by using it as a form of slide hammer. If this tool is not readily available it can, of course, be fabricated from a piece of metal tubing of the appropriate dimensions. Care should be taken however, to ensure that the end of the tube that makes contact with the seal is properly chamfered, free of burrs and absolutely square to the fork stanchion. Alternatively, a suitable socket may be used to drive the seal into position before the stanchion is fitted. Carry out a final check to ensure that the seal has been driven home squarely before retaining it in position with the spring clip.

19 Refit the dust excluder over the fork lower leg and support the fork leg in a vertical position. Replenish the fork leg with 90 cc (3.17 Imp oz) of SAE 10W/20 oil, pouring the oil into the top of the stanchion.

20 Replace the fork spring(s) in the leg (and the seating ring, where fitted). On models which have a single spring fitted to each fork leg, ensure that the close-pitch end of the spring is nearest the top of the fork leg. Refit and tighten the spring retaining plug. In practice, it was found to be difficult to locate this plug on the thread in the inside of the fork stanchion, owing to the spring pressure acting against it. This problem was eventually overcome by adapting a socket as shown in the photograph accompanying this test. A socket which is a close fit inside the bore of the stanchion will serve to keep the plug square to the thread whilst giving some means with which to bear down against the pressure of the spring. With the plug fully tightened, the fork leg is ready to be refitted to the machine.

21 Before refitting the fork legs, however, it is worth pausing to consider the advantages of fitting protective gaiters over the exposed section of fork stanchion between the lower yoke and the dust excluder. It is a proven fact that the life of the oil seal can be lengthened considerably by doing this, with the additional advantage that the surface of the fork stanchion is also protected. Several manufacturers of motorcycle accessories provide gaiters which can be adapted to fit the type of fork legs dealt with in this Section.

3.2a Take care when removing the fork spring retaining plug

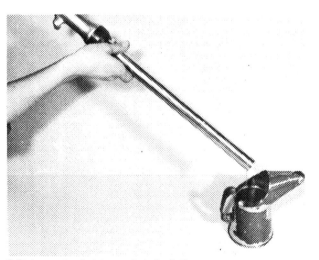

3.2b Invert the fork leg to drain it of oil

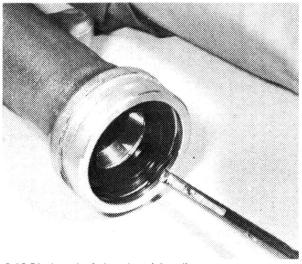

3.10 Displace the fork seal retaining clip

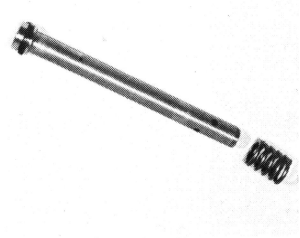

3.14 Examine the damper rod assembly

3.15a The dust excluder must be undamaged

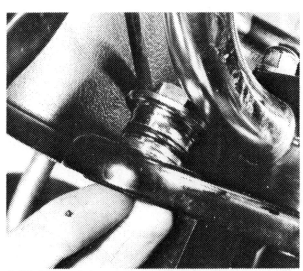

3.15b The cap bolt can only be removed for renewal of the O-ring by first detaching the handlebars

3.17a Check that the damper rod piston ring is correctly fitted

3.17b ... before inserting the rod into the fork stanchion

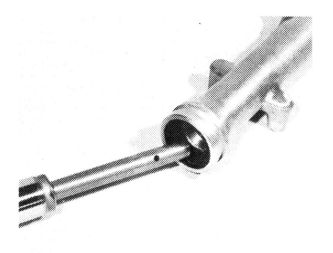

3.17c Fit both the stanchion and damper rod into the fork lower leg ...

3.17d ... place a serviceable sealing washer on the retaining bolt

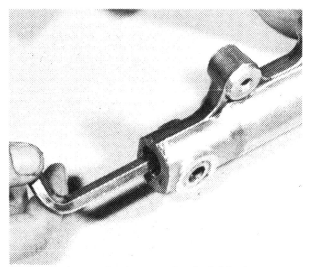

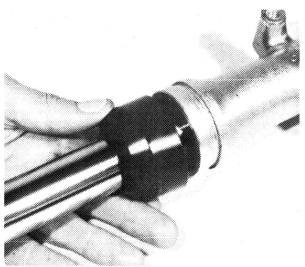

3.17e ... and secure the damper rod to the fork leg base

3.19a Slide the dust excluder into position ...

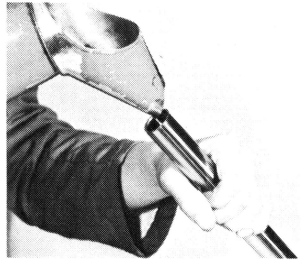

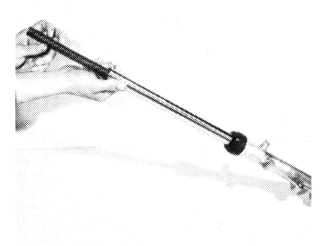

3.19b ... before replenishing the fork leg with oil ...

3.20a ... and refitting the fork spring(s)

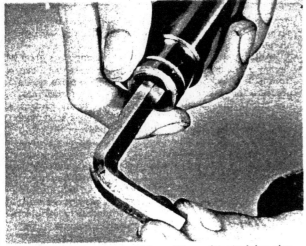

3.20b Use a socket to aid refitting of the spring retaining plug

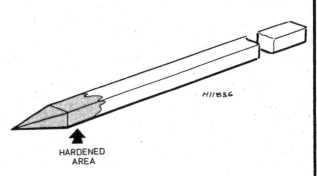

HII836

HARDENED
AREA

Fig. 4.1 Fabricated tool for holding the damper rod in
position

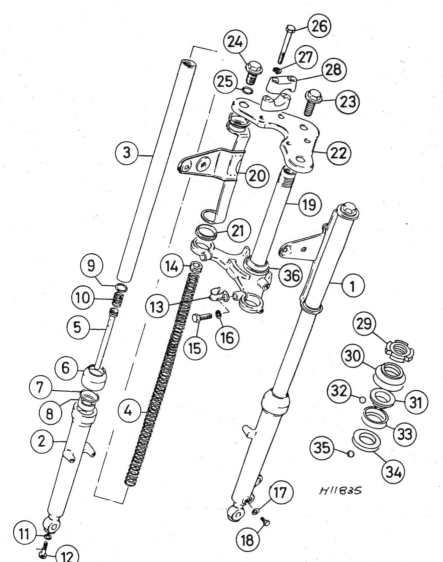

1 Complete fork leg
2 Lower leg
3 Stanchion
4 Spring
5 Damper rod
6 Dust excluder
7 Spring clip
8 Oil seal
9 Piston ring
10 Rebound spring
11 Sealing washer
12 Allen bolt
13 Cable guide
14 Fork spring retaining plug
15 Bolt – 2 off
16 Spring washer – 2 off
17 Sealing washer*
18 Drain plug*
19 Steering stem/lower yoke
20 Headlamp bracket
21 Damping ring – 4 off
22 Fork upper yoke
23 Crown bolt
24 Cap bolt
25 O-ring
26 Bolt – 2 off
27 Spring washer – 2 off
28 Handlebar clamp – 2 off
29 Adjusting nut
30 Dust cover
31 Upper bearing cone
32 Upper bearing balls
33 Upper bearing cup
34 Lower bearing cup
35 Lower bearing balls
36 Lower bearing cone
*early GP125 C models only

HII835

Fig. 4.2 Front forks

## 4  Steering head assembly : removal and refitting

1    The steering head will rarely require attention unless it becomes necessary to renew the bearings or if accident damage has been sustained. It is theoretically possible to remove the lower yoke together with the fork legs, but as this entails a considerable amount of unwieldy manoeuvring this approach is not recommended. A possible exception may arise if the fork stanchions have been damaged in an accident and are jammed in the lower yoke, and in this case a combination of this Section and Section 2 must be applied.

2    Remove the front wheel, brake caliper (where necessary), front mudguard, handlebars and fork legs as described in Section 2 of this Chapter.

3    Before disconnecting any electrical leads from the multitude of electrical components mounted on the headstock assembly, it is advisable to isolate the battery from the machine's electrical system by removing one of the leads from its battery terminal. This will safeguard against any danger of components within the electrical system becoming damaged by short circuiting of the exposed connector ends

4    Disconnect the speedometer and tachometer drive cables from the base of their respective heads. Remove the speedometer cable from the machine and thread the tachometer cable clear of the steering head assembly. The instrument console may now be freed from its mountings on the upper yoke by removing the two mounting bolts. Move the console forward away from the yoke as far as the electrical leads will allow and rest it on the headlamp assembly.

5    The procedure from this point onwards must depend on individual circumstances. For obvious reasons, the full dismantling sequence is described here, but it is quite in order to avoid as much of the dismantling as possible by careful manoeuvring of the ancillary components. Obviously, much depends on a commonsense approach and a measure of ingenuity on the part of the owner.

6    Remove the large chromium plated bolt from its location through the centre of the upper yoke. Using a soft-faced hammer, give the upper yoke a gentle tap to free it from the steering head and lift it from position. With the yoke thus removed, each of the headlamp retaining brackets can be detached from the lower yoke and the complete headlamp assembly then moved forward to clear the steering head. Carry out a final check around the lower yoke to ensure that the hydraulic brake hose (where fitted), the tachometer drive cable and any electrical leads have all been freed from their retaining clips. Finally, note the fitted position of the two electrical leads that connect to the horn and then disconnect them.

7    Support the weight of the lower yoke and, using a C-spanner of the correct size, remove the steering head bearing adjusting ring. If a C-spanner is not available, a soft metal drift may be used in conjunction with a hammer to slacken the ring.

8    Remove the dust excluder and the cone of the upper bearing. The lower yoke, complete with steering stem, can now be lowered from position. Ensure that any balls that fall from the bearings as the bearing races separate are caught and retained. It is quite likely that only the balls from the lower bearing will drop free, since those of the upper bearing will remain seated in the bearing cup. Full details of examining and renovating the steering head bearings are given in Section 5 of this Chapter.

9    Fitting of the steering head assembly is a direct reversal of that procedure used for removal, whilst taking into account the following points. It is advisable to position all eighteen balls of the lower bearing around the bearing cone before inserting the steering stem fully into the steering head. Retain these balls in position with grease of the recommended type and fill both bearing cups with the same type of grease.

10   With the lower yoke pressed fully home into the steering head, place the twenty two balls into the upper bearing cup and fit the bearing cone followed by the dust excluder. Refit the adjusting ring and tighten it, finger-tight. The ring should now be tightened until resistance to movement is felt and then

loosened $\frac{1}{8}$ to $\frac{1}{4}$ of a turn. Move the lower yoke from side to side several times and then check tighten the ring by using the aforementioned procedure.

11   Adjustment of the steering head bearings is correct when all play in the bearings is taken up but the yoke will move freely from lock to lock. Note that it is possible to place several tons pressure on the steering head bearings if they are over-tightened. The usual symptom of overtight bearings is a tendency for the machine to roll at low speeds even though the handlebars may appear to turn quite freely.

12   Finally, whilst refitting and reconnecting all disturbed components, take care to ensure that all control cables, drive cables, electrocal leads, etc are correctly routed and that proper reference is made to the list of torque wrench settings given in the Specifications Section of this Chapter and of Chapter 5. Check that the headlamp beam height has not been disturbed and ensure that all controls and instruments function correctly before taking the machine on the public highway.

## 5  Steering head bearings : examination and renovation

1    Before commencing reassembly of the steering head component parts, take care to examine each of the steering head bearing races. The ball bearing tracks of their respective cup and cone bearings should be polished and free from any indentations or cracks. If wear or damage is evident, then the cups and cones must be renewed as complete sets.

2    Carefully clean and examine the balls contained in each bearing assembly. These should also be polished and show no signs of surface cracks or blemishes. If any one ball is found to be defective, then the complete set should be renewed. Remember that a complete set of these balls is relatively cheap and it is not worth the risk of refitting items that are in doubtful condition.

3    Twenty two balls are fitted in the top bearing race and eighteen in the lower. This arrangement will leave a gap between any two balls but an extra ball must not be fitted, otherwise the balls will press against each other thereby accelerating wear and causing the steering action to be stiff.

4    The bearing outer races are a drive fit in the steering head and may be removed by passing a long drift through the inner bore of the steering head and drifting out the defective item from the opposite end. The drift must be moved progressively around the race to ensure that it leaves the steering head evenly and squarely.

5    The lower of the two inner races fits over the steering stem and may be removed by carefully drifting it up the length of the stem with a flat-ended chisel or a similar tool. Again, take care to ensure that the race is kept square to the stem.

6    Fitting of the new bearing races is a straightforward procedure whilst taking note of the following points. Ensure that the race locations within the steering head are clean and free of rust; the same applies to the steering stem. Lightly grease the stem and head locations to aid fitting of the races and drift each race into position whilst keeping it square to its location. Fitting of the outer races into the steering head will be made easier if the opposite end of the head to which the race is being fitted has a wooden block placed against it to absorb some of the shock as the drift strikes the race.

## 6  Fork yokes : examination

1    To check the top yoke for accident damage, push the fork stanchions through the bottom yoke and fit the top yoke. If it lines up, it can be asumed the yokes are not bent. Both yokes must also be checked for cracks. If they are damaged or cracked. fit serviceable replacements.

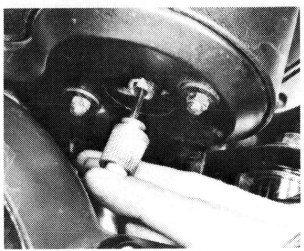

4.4a Disconnect each instrument drive cable ...

4.4b ... and remove the instrument console mounting bolts

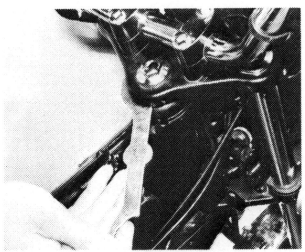

4.10 Do not overtighten the steering head bearings

### 7 Swinging arm fork : removal, examination, renovation and fitting

1    The rear fork of the frame is of the swinging arm type. This unit is of tubular steel construction and pivots on a shaft which passes through each of the fork legs and through both sides of the main frame assembly. Two bonded rubber bushes provide an efficient bearing unit, each bush being a press fit into the end of each fork leg. The pivot shaft is retained in position by a locknut with a plain washer.

2    Any wear in the pivot bearings of the swinging arm will cause imprecise handling of the machine, with a tendency for the rear end of the machine to twitch or hop. Worn bearings can be detected by placing the machine on its centre stand and then pulling and pushing vigorously on the rearmost end of one of the fork legs whilst holding on to a point on the main frame with the other hand. Any play in the bearings will be magnified by the leverage effect produced.

3    When wear develops in the bearings of the swinging arm, necessitating their renewal, the renovation procedure is quite straightforward. Commence by removing the rear wheel as described in Section 16 of Chapter 5.

4    The final drive chain guard is secured to the swinging arm by two bolts. Removal of the guard is not strictly necessary, although it will facilitate swinging arm detachment.

5    Remove the cap nut and plain washer from each of the suspension unit to swinging arm fork attachment points. Grasp the lower section of each suspension unit and pull it outwards until it clears the fork leg. Swing the fork down and rest its leg on an area of padding, a wooden block is ideal. Leave the suspension units hanging from the frame but slacken their frame mounting nuts so that each unit is free to move. This will make reattachment of each unit to the swinging arm fork a great deal easier.

6    The pivot shaft retaining nut will have been removed, along with its plain washer, during removal of the exhaust system in order to facilitate withdrawal of the rear wheel spindle. Support the crossmember of the swinging arm and draw the pivot shaft out of position. If the shaft proves to be stubborn, use a hammer and soft metal drift to displace it. Manoeuvre the swinging arm rearwards so that it clears the frame mounting points and final drive chain and then lift it clear of the machine. Give each component part a thorough clean before commencing the following examination and renovation procedures.

7    Check the pivot shaft for wear. If the shank of the shaft is seen to be stepped or badly scored, then it must be renewed. Remove all traces of corrosion and hardened grease from the shaft before checking it for straightness by rolling it on a flat surface, such as a sheet of plate glass, whilst attempting to insert a feeler gauge of 0.6 mm (0.02 in) thickness beneath it. Alternatively, place the shaft on two V-blocks and measure the amount of runout on its shank with a dial gauge. If the amount of runout measured exceeds 0.6 mm (0.02 in), replace the shaft with a straight item. Note that a bent pivot shaft will prevent the swinging arm fork from moving smoothly about its axis.

8    Detach the nylon buffer from the left-hand bearing housing and inspect it for signs of excessive wear. If it is considered that the buffer no longer serves as an adequate means of protecting the metal of the bearing housing against interference from the final drive chain, then it must be renewed.

9    Carefully inspect the structure of the swinging arm fork for signs of distortion, failure or any other damage which may lead to eventual failure of the component. It is worth taking steps to remove any corrosion from areas where the paint finish has been eroded away and then reprotecting the bared surface. Pay particular attention to the welds between component parts. If distortion is suspected, return the component to a Suzuki service agent who will be able to confirm whether or not replacement is necessary.

10    In order to renew each of the bonded rubber bushes, it is first necessary to fabricate a tool with which to press each bush

out of its housing. It is suggested that a tool similar to the one shown in the accompanying diagram is made, utilising a short length of thick-walled tube, the inside diameter of which is slightly larger than the outside diameter of the bush, a high tensile bolt and nut, and two thick plate washers, one of which has an outer diameter slightly smaller than that of the bush. Note that, as a means of removal, attempting to drive the bushes out will probably prove unsuccessful, because the rubber will effectively damp out the driving force, and damage to the housing lugs may occur.

11  If, due to corrosion between the mating faces of the bush and swinging arm lug, the bushes are reluctant to move, even using this method, it is recommended that the unit be returned to a Suzuki service agent whose expertise can be brought to bear on the problem.

12  New bushes may be driven in using a tubular drift against the outer sleeve, or by reversing the removal operation, using the fabricated puller. Whichever method is adopted, the outer sleeve should be lubricated sparingly, and care must be taken to ensure that the bush remains square with the housing bore. The swinging arm must be properly supported whilst doing this.

13  Reassemble and refit the swinging arm by reversing the removal and dismantling procedures. Lightly grease the shank of the pivot shaft before inserting it into position, this will prevent it from becoming corroded and thus seizing to each of

the bushes. Check that the fork pivots smoothly about its axis before reconnecting the suspension units and tightening their retaining nuts to a torque loading of 2.0 -3.0 kgf m (14.5 - 21.5 lbf ft). Remember, upon refitting the exhaust system, that the pivot shaft retaining nut must be tightened to a torque loading of 4.5 - 7.0 kgf m (32.5 - 50.5 lbf ft).

## 8  Rear suspension units : examination, adjustment, removal and fitting

1  Rear suspension units of the hydraulically damped type are fitted to the machines used in this Manual. Each unit comprises a hydraulic damper, effective primarily on rebound, and a concentric spring. It is secured to the frame and swinging arm by means of rubber-bushed lugs at each end of the unit. The units are provided with an adjustment of the spring tension, giving five settings. The settings can be easily altered without moving the units from the machine by using a tommy bar in the hole directly below the springs. Turning clockwise will increase the spring tension and stiffen the rear suspension, turning anti-clockwise will lessen the spring tension and therefore soften the ride. As a general guide the softest setting is recommended for road use only, when no pillion passenger is carried. The hardest setting should be used when a heavy load is carried, and during

7.5a Note the position of the retaining washer ...

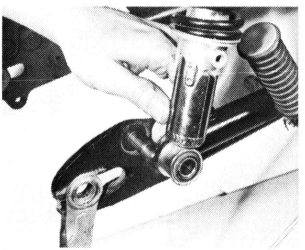

7.5b ... before pushing each suspension unit clear of the swinging arm fork

7.6 The swinging arm fork assembly

8.1 Alter the suspension setting by turning the adjuster

high-speed riding. The intermediate positions may be used as conditions dictate.

2　There is no means of draining the units or topping up, because the dampers are built as a sealed unit. If the damping fails or if the units start to leak, the complete damper assembly must be renewed. This applies equally if the damper rod has become bent, or its chromed surface badly corroded or damaged. Check also for deterioration of the rubber mounting bushes and of the rubber buffer within the spring. The piston housing must be free of damage if it is to function correctly.

3　If necessary, the suspension units can be removed from the frame and swinging arm attachment studs simply by removing the upper and lower cap nuts and plain washers and pulling the unit out away from the machine.

4　When fitting a suspension unit, employ a procedure which is a reversal of that used when removing the unit, whilst noting the following points. Before refitting the original or any seal units, take the opportunity to give them a thorough clean. Do not, under any circumstances, grease the chromed surface of the damper rod; any break down of this chromed surface will quickly lead to failure of the unit seal, thereby rendering the unit ineffective. Check that all the plain washers are correctly located before fitting and tightening the cap nuts to a torque loading of 2.0 - 3.0 kgf m (14.5 - 21.5 lbf ft).

5　Note that in the interests of good roadholding, it is essential that both suspension units have the same load setting. If renewal is necessary, the units must be replaced as a matched pair.

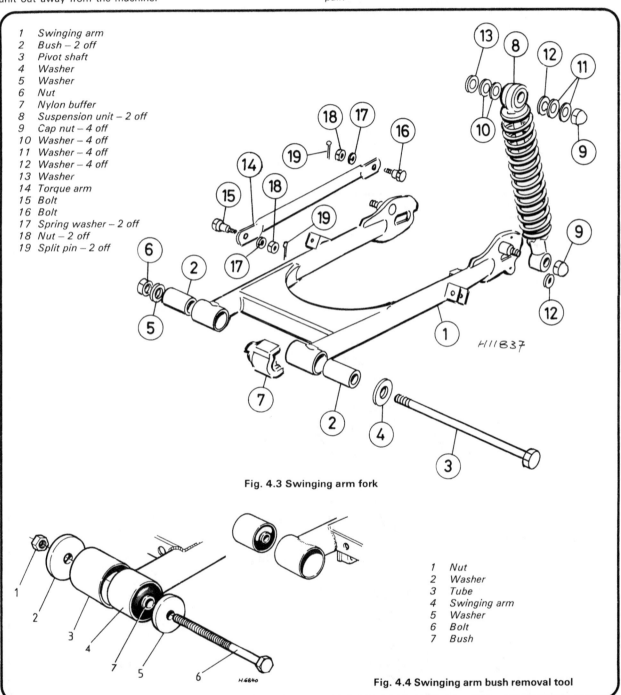

1　Swinging arm
2　Bush – 2 off
3　Pivot shaft
4　Washer
5　Washer
6　Nut
7　Nylon buffer
8　Suspension unit – 2 off
9　Cap nut – 4 off
10　Washer – 4 off
11　Washer – 4 off
12　Washer – 4 off
13　Washer
14　Torque arm
15　Bolt
16　Bolt
17　Spring washer – 2 off
18　Nut – 2 off
19　Split pin – 2 off

Fig. 4.3 Swinging arm fork

1　Nut
2　Washer
3　Tube
4　Swinging arm
5　Washer
6　Bolt
7　Bush

Fig. 4.4 Swinging arm bush removal tool

## 9  Frame assembly : examination and renovation

1   If the machine is stripped for a complete overhaul, this affords a good opportunity to inspect the frame for cracks or other damage which may have occurred in service. Check the points at which the front downtube and the frame top tubes join the steering head; these are the points where fractures are most likely to occur. The straightness of the tubes concerned will show whether the machine has been involved in a previous accident.

2   Check carefully areas where corrosion has occurred on the frame. Corrosion can cause a reduction in the material thickness and should be removed by use of a wire brush and derusting agents. After the machine has covered a considerable mileage, it is advisable to examine the frame closely for signs of cracking or splitting at the welded joints.

3   If the frame is broken or bent, professional attention is required. Repairs of this nature should be entrusted to a competent repair specialist, who will have available all the necessary jigs and mandrels to preserve correct alignment. Repair work of this nature can prove expensive and it is always worthwhile checking whether a good replacement frame of identical type can be obtained at a reasonable cost.

4   Remember that a frame which is in any way damaged or out of alignment will cause, at the very least, handling problems. Complete failure of a main frame component could well lead to a serious accident.

## 10  Centre stand : exmination and servicing

1   The centre stand is retained between the two engine rear mounting plates by a single bolt which acts as a pivot and which is retained in position by a spring washer and nut. A bush is fitted between the cross tube of the centre stand and the pivot bolt and acts as a bearing surface between the two components.

2   The pivot assembly on any centre stand is often neglected with regard to lubrication and this eventually leads to excessive wear. It is prudent to detach the centre stand from the machine from time to time in order to allow grease to be applied to the bearing surfaces of the bush, once it has been withdrawn from its location in the stand cross tube.

3   Inspect the shank of the pivot bolt for signs of its being bent or worn and renew it, if necessary. Carry out a similar inspection on the cross tube of the stand. Before relocating the spring washer over the pivot bolt, check that it has not become flattened. If this is the case, then it must be renewed. If the pivot bolt retaining nut works loose even when fastened correctly to the specified torque setting, substitute a Nyloc-insert self-locking nut, available from any official Suzuki service agent; do not merely overtighten the nut to secure it, as this will distort the supporting lugs and jam the stand.

4   Check that the holes in the spring to frame pivot plate have not become elongated. This will only occur after the machine has been in use for a considerable amount of time. Elongation of either one of these holes will effectively reduce the tension in the spring whilst the stand is in the retracted position, with the resulting effect that the stand will not be drawn tight up against its stop.

5   Check that the return spring is free from fatigue and in good condition. Smearing grease along the length of the spring will lessen the chances of its becoming corroded. Remember that a broken or weak spring will cause the stand to fall whilst the machine is in motion, with the resulting danger that once it catches on some obstacle on the road surface, the balance of the machine will be drastically affected.

## 11  Prop stand : examination and servicing

1   The prop stand bolts to a lug which forms part of the front footrest bracket. An extension spring is fitted to ensure that the stand retracts when the weight of the machine is taken off the

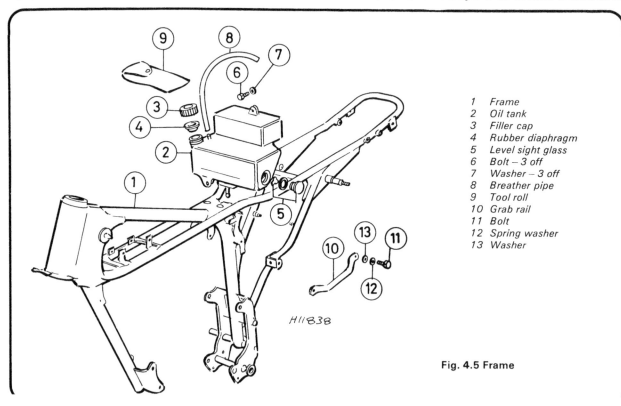

1    Frame
2    Oil tank
3    Filler cap
4    Rubber diaphragm
5    Level sight glass
6    Bolt – 3 off
7    Washer – 3 off
8    Breather pipe
9    Tool roll
10   Grab rail
11   Bolt
12   Spring washer
13   Washer

H1183B

**Fig. 4.5 Frame**

stand and the stand is kicked rearwards. When properly retracted, the stand should be tight against its stop and well out of the way.

2    Check to ensure that the nut which serves to retain the bolt about which the stand pivots is fully tightened and that the pivot surfaces are well lubricated with either grease or motor oil. The extension spring must be free from fatigue and in good condition. Smearing grease along the length of the spring will lessen the chances of its becoming corroded.

3    Finally, remember that a serious accident is almost inevitable if the stand extends whilst the machine is in motion.

4    Later GP 100 UD and all UL models are fitted with a flyback-type prop stand which has two springs attached so that it will always retract itself immediately the machine's weight is removed from it. Servicing procedures are as described above.

### 12 Footrests : examination and renovation

1    The front footrest assembly comprises a metal bracket with removable rubber pads. This bracket passes beneath the rear of the engine unit and is secured to each of the two engine

mounting plates by means of a bolt with a spring washer beneath its head. If this bracket is bent in a spill or through the machine falling over, it can be removed from the machine and straightened in a vice whilst being heated to a dull red with a blow lamp or welding torch.

2    The pillion footrests are pivoted on clevis pins. If an accident occurs, it is probable that the footrest peg will move and remain undamaged. A bent peg may be detached from the mounting, after removing the clevis pin securing split-pin and the clevis pin itself. The damaged peg can be straightened in a vice, using a blowlamp flame to apply heat at the area where the bend occurs. The footrest rubber will, of course, have to be removed as the heat will render it unfit for service.

3    Note, with both types of footrest, that if there is evidence of failure of the metal either before or after straightening, it is advised that the damaged component is renewed. If a footrest breaks, loss of machine control is almost inevitable.

4    When refitting the front footrest bracket, ensure that the spring washers are both serviceable and that both bolts are fully tightened. When refitting each of the pillion footrests, ensure that a new split-pin is used to retain the clevis pin in position and that the legs of the pin are correctly spread.

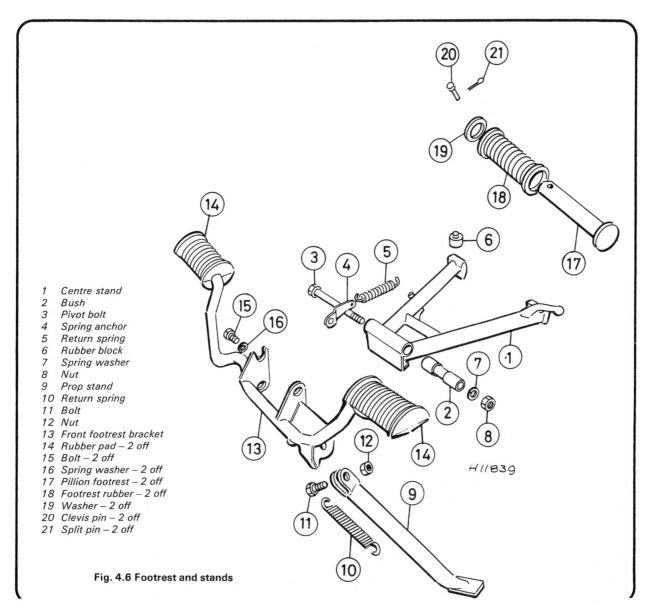

1    Centre stand
2    Bush
3    Pivot bolt
4    Spring anchor
5    Return spring
6    Rubber block
7    Spring washer
8    Nut
9    Prop stand
10   Return spring
11   Bolt
12   Nut
13   Front footrest bracket
14   Rubber pad – 2 off
15   Bolt – 2 off
16   Spring washer – 2 off
17   Pillion footrest – 2 off
18   Footrest rubber – 2 off
19   Washer – 2 off
20   Clevis pin – 2 off
21   Split pin – 2 off

H11839

**Fig. 4.6 Footrest and stands**

### 13 Rear brake pedal : examination and renovation

1 The rear brake pedal is mounted in a pivot pin which forms part of the engine rear mounting plate assembly and is located on the right-hand side of the machine. The pedal is retained on the pivot pin by means of a plain washer and a split-pin. Full access to the pin can only be gained if the exhaust system is detached from the machine as detailed in Section 17 of Chapter 2. In the event of damage occurring, the pedal should be removed from the machine and treated similarly to a bent footrest as described in Section 12 of this Chapter. Note that the warning relating to footrest breakage applies equally to the brake pedal because it follows that failure is most likely to occur when the brake is applied firmly, which is when it is required most.

2 To remove the pedal from its pivot, detach the rear stop lamp switch operating spring and then remove the split-pin and plain washer from the pivot pin. Move the pedal outboard so that access can be gained to the split-pin and plain washer which serve to retain the brake operating rod to the pedal. Detach the rod from the pedal and remove the pedal from the machine together with its return spring.

3 Inspect the pedal return spring for signs of fatigue or failure and renew it if necessary. It is advisable, if the brake pedal has been in any way damaged, to check the condition of the pivot pin. This is also the case if operation of the pedal is noticed to be stiff and the pedal fails to return immediately when released. Check the pin for straightness by placing a straight edge alongside it for comparison. If bent, the pin must be straightened or replaced. Do not, under any circumstances, attempt to hammer this pin back into position without first heating it to a dull red; take the utmost care to protect any engine or frame components in the immediate vicinity whilst doing this.

4 If the pedal is seen to have seized through lack of lubrication or if any of the bearing surfaces are found to be corroded, clean both the length of the pin and the bore of the pedal with fine grade emery cloth before refitting the pedal back onto the pin and checking for excessive play between the two components. If either component is thought to be excessively worn then it must be renewed.

5 Lubricate the length of the pin with a high quality lithium based grease and fit the pedal onto it. Reconnect the brake operating rod, using a new split-pin and spreading its legs to retain it in position. Push the pedal fully home onto the pin whilst checking that the return spring is correctly located and that the pedal pivots smoothly around the pin and returns to its correct position against the footrest. Lightly grease the return spring to lessen the risk of its being corroded by road salts.

6 Reconnect the stop lamp switch operating spring and retain the pedal to the pivot pin by refitting the plain washer and a new split-pin. Before riding the machine, carry out a final check to ensure that both the rear brake and the stop lamp switch are in correct adjustment and are functioning correctly.

### 14 Kickstart lever : examiantion and renovation

1 The kickstart lever is splined and is secured to its shaft by means of a pinch bolt. The kickstart crank swivels so that it can be tucked out of the way when the engine is started. It is held in position on the swivel by a washer and circlip. A spring-loaded ball bearing locates the kickstart arm in either the operating or folded position; if the action becomes sloppy it is probable that the spring behind the ball bearing needs renewing. It is advisable to remove the circlip and washer occasionally, so that the kickstart crank can be detached and the swivel greased.

2 It is unlikely that the kickstart crank will bend in an accident unless the machine is ridden with the kickstart in the operating and not folded position. It should be removed and straightened, using the same technique as that recommended for the footrests.

### 15 Steering lock : general description and renewal

1 A steering lock is attached to a lug on the steering head by means of two screws, each with a spring washer beneath its head. When in the locked position, a bar extends from the body of the lock and abuts against a projection which forms part of the casting of the lower yoke. This effectively prevents the handlebars from being turned once they are set on full lock.

2 If the lock malfunctions, then it must be renewed. A repair is impracticable. When the lock has been renewed, ensure a key which matches the lock is obtained and carried when the machine is in use.

### 16 Speedometer and tachometer heads : removal, inspection and fitting

1 The speedometer and tachometer heads are mounted together in a single console which is seated on top of the steering head assembly. Each instrument is secured in position by two dome nuts, each with a plate washer, which attach to studs protruding from the base of the console casing.

2 Each instrument may be detached from the machine by first disconnecting the drive cable from its base. Remove all four of the dome nuts, with their washers, from the base of the console and pull the baseplate down away from the console casing to expose the bulb connections in the base of the instrument. Unplug these bulb connections from the instrument and pull the instrument up out of the console casing. In the case of the speedometer head, it will be necessary to detach the tripmeter adjusting knob from the side of the instrument before the instrument can be pulled clear of the console. This can be done simply by grasping the knob and pulling it off its retaining shaft.

3 If either instrument fails to record, check the drive cable first before suspecting the head. If the instrument gives a jerky response it is probably due to a dry cable, or one that is trapped or kinked.

4 The speedometer and tachometer heads cannot be repaired by the private owner, and if a defect occurs a new instrument has to be fitted. Remember that a speedometer in correct working order is required by law on a machine in the UK and also in many other countries.

5 Refer to the following Sections of this Chapter for details of servicing the instrument drive assemblies. On completion of servicing either the instruments or their drive assemblies, refit them by using the reverse procedure to that given for removal. Check that all disturbed electrical connections are properly remade and that the drive cables are correctly routed.

### 17 Speedometer and tachometer drive cables : examination and renovation

1 It is advisable to detach the speedometer and tachometer drive cables from time to time in order to check whether they are adequately lubricated and whether the outer cables are compressed or damaged at any other point along their run. A jerky or sluggish movement at the instrument head can often be attributed to a cable fault.

2 To grease the cable, uncouple both ends and withdraw the inner cable. (On some model types this may not be possible in which case a badly seized cable will have to be renewed as a complete assembly). After removing any old grease, clean the inner cable with a petrol soaked rag and examine the cable for broken strands or other damage. Do not check the cable for broken strands by passing it through the fingers or palm of the hand, this may well cause a painful injury if a broken strand snags the skin. It is best to wrap a piece of rag around the cable and pull the cable through it, any broken strands will snag on the rag.

3   Regrease the cable with high melting point grease, taking care not to grease the last six inches closest to the instrument head. If this precaution is not observed, grease will work into the instrument and immobilise the sensitive movement.

4   If the cable breaks, it is usually possible to renew the inner cable alone, provided the outer cable is not damaged or compressed at any point along its run. Before inserting the new inner cable, it should be greased in accordance with the instructions given in the preceding paragraph. Try to avoid tight bends in the run of the cable because this will accelerate wear and make the instrument movement sluggish.

### 18 Speedometer and tachometer drives : location and examination

1   On models fitted with a front brake of the full-width drum type, the speedometer is driven from a gearbox which forms an integral part of the brake backplate assembly. Drive is transmitted through the slotted end of the inner section of hub which engages with the drive plate in the gearbox. Full details of servicing this particular unit are contained in Section 4 of the following Chapter.

2   Models fitted with a front brake of the disc and hydraulically-operated caliper type, have a speedometer gearbox fitted over the front wheel spindle on the right-hand side of the wheel hub. Drive is transmitted through slots cast in the hub centre which engage with the drive plate of the gearbox. Provided that the gearbox is detached and repacked with grease from time to time, very little wear should be experienced. In the event of failure, the complete unit should be renewed. Full details of detaching and servicing this unit are contained in Section 4 of the following Chapter.

3   The drive for the tachometer is taken from a point on the crankcase adjacent to the kickstart lever. The drive is taken from a gear which forms part of the oil pump drive assembly. Full details of removal and fitting of this gear are contained within the relevant Sections of Chapter 1. It is unlikely that this internal drive mechanism will give trouble during the normal service life of the machine, particularly since it is fully enclosed and effectively lubricated.

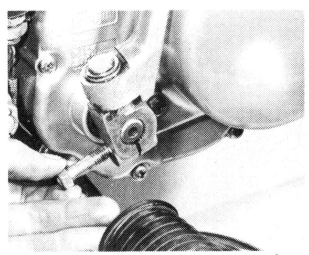

14.1 The kickstart lever is secured to its shaft by means of a pinch bolt

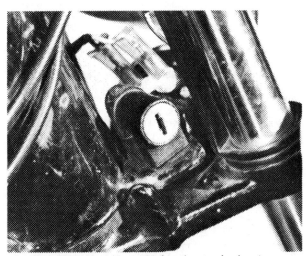

15.1 The steering lock is mounted on the steering head

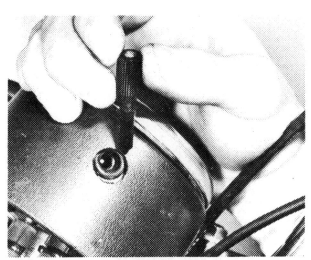

16.2a Detach the tripmeter adjusting knob ...

16.2b ... before removing the speedometer head

## 19 Dualseat : removal and refitting

1    Remove the dualseat by releasing its two mounting bolts. These are located one either side towards the rear of the seat. The seat may then be lifted up and rearwards off its forward mounting point.

2    It will be seen that both of these bolts also serve to support the rearmost section of rear mudguard. To avoid loss of the bolts, relocate them in the frame to seat attachment points whilst ensuring that each bolt has one serviceable spring washer and the required number of plain washers located beneath its head.

3    Refitting the seat is a reversal of the removal procedure. It is important that the seat is correctly engaged on the forward mounting point and that the rear mounting bolts are fully tightened. Should the seat become detached whilst the machine is in motion, the resulting loss in balance of the rider could well prove disastrous.

## 20 Cleaning the machine

1    After removing all surface dirt with a rag or sponge which is washed frequently in clean water, the machine should be allowed to dry thoroughly. Application of car polish or wax to the cycle parts will give a good finish, particularly if the machine receives this attention at regular intervals.

2    The plated parts should require only a wipe with a damp rag, but if they are badly corroded, as may occur during the winter when the roads are salted, it is permissible to use one of the proprietary chrome cleaners. These often have an oily base which will help to prevent corrosion from recurring.

3    If the engine parts are particularly oily, use a cleaning compound such as Gunk or Jizer. Apply the compound whilst the parts are dry and work it in with a brush so that it has an opportunity to penetrate and soak into the film of oil and grease. Finish off by washing down liberally, taking care that water does not enter the carburettors, air cleaners or the electrics.

4    If possible, the machine should be wiped down immediately after it has been used in the wet, so that it is not garaged under damp conditions which will promote rusting. Make sure that the chain is wiped and re-oiled, to prevent water from entering the rollers and causing harshness with an accompanying rapid rate of wear. Remember there is less chance of water entering the control cables and causing stiffness if they are lubricated regularly as described in the Routine Maintenance Section.

## 21 Cleaning the plastic mouldings

1    The moulded plastic cycle parts, which include the side-panels, the seat fairing and some of the crankcase inspection covers, require treating in a different manner to the metal cycle parts.

2    These plastic parts will not respond to cleaning in the same way as painted metal parts; their construction may be adversely affected by traditional cleaning and polishing techniques, and lead as a result to the surface finish deteriorating. It is best to wash these parts with a household detergent solution, which will remove oil and grease in a most effective manner.

3    Avoid the use of scouring powder or other abrasive cleaning agents because this will score the surface of the mouldings, making them more receptive to dirt and permanently damaging the surface finish.

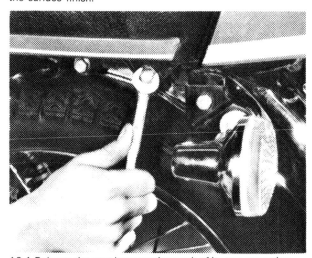

19.1 Release the seat by removing each of its two mounting bolts

## 22 Fault diagnosis : frame and forks

| Symptom | Cause | Remedy |
|---|---|---|
| Machine veers to the left or right with hands off handlebars | Wheels out of alignment<br>Forks twisted<br>Frame bent | Check and realign<br>Strip and repair<br>Strip and repair or renew |
| Machine tends to roll at low speeds | Steering head bearings not adjusted correctly or worn | Check adjustment and renew bearings if necessary |
| Machine tends to wander | Worn swinging arm bearings | Check and renew bearings |
| Forks judder when front brake applied | Steering head bearings slack<br>Fork components worn | Adjust bearings. Strip forks, and renew all worn parts |
| Forks bottom | Short of oil | Replenish with correct viscosity oil |
| Fork action stiff | Fork legs out of alignment<br>Bent shafts, or twisted ie, yokes | Slacken clamp bolts, front wheel spindle and top bolts. Pump forks several times and tighten from bottom upwards. Strip and renew parts, if undamaged |
| Machine pitches badly | Defective rear suspension units or ineffective fork damping | Check damping action<br>Check grade and quantity of oil in front forks |

# Chapter 5  Wheels, brakes and tyres

## Contents

## Specifications

### Wheels

| | | |
|---|---|---|
| Type: | | |
|     GP100 ED models .................................................. | | Cast alloy, Suzuki ten-spoke design |
|     All other models ................................................... | | Conventional, steel spoked with chromed steel rim |
| Size: | | |
|     Front and rear ..................................................... | | 18 inch |
| Rim runout: | | |
|     Radial and axial service limit ............................... | | 2.0 mm (0.08 in) |
| Spindle runout: | | |
|     Front and rear service limit ................................. | | 0.25 mm (0.010 in) |

### Brakes

| | |
|---|---|
| Front: | |
|     Type: | |
|         Suffix U models ................................................. | Internally expanding, single leading shoe, drum |
|         All other models ............................................... | Single hydraulic disc |
|     Fluid specification ............................................... | SAE J1703 |
|     Brake shoe lining thickness service limit .......................... | 1.5 mm (0.06 in) |
|     Brake drum internal diameter .............................................. | 130 mm (5.12 in) |
|     Service limit ....................................................................... | 130.7 mm (5.15 in) |
|     Brake pad wear limit .......................................................... | To red limit line |
|     Brake disc thickness .......................................................... | 3.8 – 4.2 mm (0.150 – 0.165 in) |
|     Service limit ....................................................................... | 3.0 mm (0.12 in) |
|     Brake disc runout service limit ........................................... | 0.3 mm (0.01 in) |
|     Caliper cylinder bore .......................................................... | 33.960 – 34.0 mm (1.3370 – 1.3386 in) |
|     Caliper piston diameter ....................................................... | 33.890 – 33.910 mm (1.3342 – 1.3350 in) |
|     Master cylinder bore ........................................................... | 12.700 – 12.743 mm (0.50 – 0.5017 in) |
|     Master cylinder piston diameter .......................................... | 12.650 – 12.670 mm (0.4980 – 0.4988 in) |

Rear:
    Type ................................................................... Internally expanding, single leading shoe, drum
    Brake shoe lining thickness service limit ........................ 1.5 mm (0.06 in)
    Brake drum internal diameter:
        GP100 models .............................................. 110 mm (4.33 in)
        GP125 models .............................................. 130 mm (5.12 in)
    Service limit:
        GP100 models .............................................. 110.7 mm (4.36 in)
        GP125 models .............................................. 130.7 mm (5.15 in)

## Tyres

| | GP100 | GP125 |
|---|---|---|
| Size: | | |
|   Front | 2.50 – 18 – 4PR | 2.75 – 18 – 4PR |
|   Rear | 2.75 – 18 – 4PR | 3.00 – 18 – 4PR |
| Pressures: | | |
|   Solo: | | |
|     Front | 25 psi (1.75 kg/cm²) | 25 psi (1.75 kg/cm²) |
|     Rear | 28 psi (2.0 kg/cm²) | 32 psi (2.25 kg/cm²) |
|   With pillion: | | |
|     Front | 25 psi (1.75 kg/cm²) | 25 psi (1.75 kg/cm²) |
|     Rear | 32 psi (2.25 kg/cm²) | 36 psi (2.50 kg/cm²) |
| Manufacturer's recommended minimum tread depth | 1.6 mm (0.06 in) | 1.6 mm (0.06 in) |

## Torque wrench settings

| | kgf m | lbf ft |
|---|---|---|
| Front wheel spindle nut | 2.7 – 4.3 | 19.5 – 31.0 |
| Rear wheel spindle nut | 3.6 – 5.2 | 26.0 – 37.5 |
| Rear torque arm nuts | 1.0 – 1.5 | 7.0 – 11.0 |
| Front brake caliper spindle bolts | 1.5 – 2.0 | 11.0 – 14.5 |
| Front brake caliper mounting to fork leg bolts | 1.5 – 2.5 | 11.0 – 18.0 |
| Front brake master cylinder clamp bolts | 0.6 – 0.9 | 4.5 – 6.5 |
| Hydraulic hose union bolts | 2.5 – 4.0 | 18.0 – 29.0 |
| Brake cam lever nut (front and rear) | 0.5 – 0.8 | 3.5 – 6.0 |

## 1  General description

The design of wheel fitted to the machines covered in this Manual varies between the model types. At the time of writing, the traditional design of a chromed steel rim laced to an alloy hub by steel spokes, is fitted to all GP100 and GP125 models except the GP100 ED model which is fitted with a cast alloy wheel of Suzuki's own ten-spoke design. Both types of wheel utilise conventional tubed tyres, the sizes of which are given in the Specifications Section of this Chapter.

GP100 models with the letter suffix U have a front brake of standard cable operated, single leading shoe, full-width hub design. All other models covered in this Manual are equipped with a hydraulically operated, single disc, front brake which utilises a single piston type of caliper. All models utilise the same type of rod operated rear brake, this being similar in design to the front drum brake fitted to the GP100 U models.

## 2  Front wheel: examination and renovation

1  Place the machine on its centre stand and position blocks beneath the engine crankcase so that the front wheel is raised clear of the ground. Spin the wheel and check the rim alignment by means of a pointer or dial gauge set against the rim. The rim should be no more than 2.0 mm (0.08 in) out of true in either its radial or axial planes.
2  On machines fitted with conventional wire-spoked wheels, small irregularities in alignment can be corrected by tightening the spokes in the affected area, although a certain amount of experience is necessary to prevent over-correction. Any flats in the wheel rim will be evident at the same time. These are more difficult to remove and in most cases it will be necessary to have the wheel rebuilt on a new rim. Apart from the effect on stability, a flat will expose the tyre bead and walls to greater risk of damage if the machine is run with a deformed wheel.
3  Check also for loosen and broken spokes. Tapping the spokes is the best guide to tension. A loose spoke will produce a quite different sound and should be tightened by turning the nipple in an anti-clockwise direction. Always check for runout by spinning the wheel again. If the spokes have to be tightened by an excessive amount, it is advisable to remove the tyre and tube as detailed in Section 21 of this Chapter. This will enable the protruding ends of the spokes to be ground off, thus preventing them from chafing the inner tube and causing punctures.
4  On machines fitted th cast alloy wheels, Suzuki recommend that a wheel which is more than the specified 2.0 mm (0.08 in) out of alignment should be renewed. This is, however, a counsel of perfection; a run out somewhat greater than this can probably be accommodated without noticeable effect on steering. No means is available for straightening a warped wheel without resorting to the expense of having the wheel skimmed on all faces. If warpage was caused by impact during an accident, the safest measure is to renew the wheel complete.
5  When inspecting a cast alloy wheel, carefully check the complete wheels for cracks and chipping, particularly at the spoke roots and the edge of the rim. As a general rule a damaged wheel must be renewed as cracks will cause stress points which may lead to sudden failure under heavy load.

Small nicks may be radiused carefully with a fine file and emery paper (No 600 – No 1000) to relieve the stress. If there is any doubt as to the condition of a wheel, advice should be sought from a Suzuki repair specialist.

6    Each cast alloy wheel is covered with a coating of lacquer to prevent corrosion. If damage occurs to the wheel and the lacquer finish is penetrated, the bared aluminium alloy will soon start to corrode. A whitish grey oxide will form over the damaged area, which in itself is a protective coating. This deposit however, should be removed carefully as soon as possible and new protective coating of lacquer applied.

7    Note that on both wheel types, it is possible that worn wheel bearings may cause rim run out. Worn bearings should be renewed by following the procedure described in Section 5 of this Chapter.

## 3    Front wheel: removal and fitting

1    Place the machine on the centre stand so that it is resting securely on firm ground with the front wheel well clear of the ground. If necessary place wooden blocks below the crankcase to raise the wheel.

2    Displace the split-pin from the wheel spindle retaining nut and remove the nut with its plain washer. Disconnect the speedometer cable from the speedometer gearbox by unscrewing its knurled retaining ring and pulling it out of its location. Allow the cable to hang clear of the wheel. On machines equipped with a drum brake, disconnect the brake operating cable from the cam shaft operating arm and the brake backplate by removing the nut from the threaded end of the cable and pulling the cable clear of the wheel. Refit the rubber dust seal, the spring and the trunnion from the operating arm and the nut to the cable to prevent loss and allow the cable to hang clear of the wheel.

3    Fit a tommy bar through the hole provided in the end of the wheel spindle, support the wheel and pull the spindle clear of the fork legs. Carefully lower the wheel and then manoeuvre it clear of the machine.

4    On machines equipped with a disc brake, remove the speedometer gearbox and place it to one side. Full details of servicing this assembly are given in Section 4 of this Chapter. Great care must be taken with this type of braking system not to operate the brake lever once the brake disc is removed from between the brake pads since fluid pressure may displace the piston and cause fluid leakage. Additionally, the distance betwen the pads will be reduced thereby making refitting of the

wheel extremely difficult. To prevent any chance of this happening, it is a good idea to place a hardwood wedge between the two pads directly the wheel is removed.

5    Fit the wheel by reversing the procedure used for removal whilst noting the following points.

### Disc brake models

6    With this type of brake, always ensure that the brake disc is free of grease or oil contamination before commencing the fitting operation. Ensure that the speedometer gearbox has been correctly serviced before relocating it in the wheel hub. Note that the drive tangs of the gearbox must fit in the grooves within the wheel hub and that the embossed arrow-mark on the gearbox casing must point upwards.

7    Take care, when lifting the wheel into position btween the fork legs, to ensure that the brake disc enters between the brake pads cleanly, otherwise damage to the pads will result.

### Drum brake models

8    The most important point to note when fitting a wheel equipped with a drum brake is to ensure that the spigot which forms part of the fork lower leg engages correctly in the slot cast in the brake backplate. Note that if the brake backplate is allowed to rotate, due to its not being engaged correctly with the fork leg, the wheel will lock on the first application of the brake, with disastrous consequences.

9    Before reconnecting the brake operating cable, inspect the rubber dust seal for signs of splitting or perishing and renew it, if considered necessary. The return spring also should be renewed if it is seen to be badly corroded, fatigued or broken. Once the cable is connected, check it for correct adjustment and operation by following the procedure given in Section 12 of this Chapter.

### All models

10    Before inserting the wheel spindle, apply a light film of grease along its length. Align the wheel between the fork legs and push the spindle into position, giving it a light tap with a soft-faced hammer to seat it. Fit the spindle retaining nut, with its plain washer, and tighten it to a torque setting of 2.7 – 4.3 kgf m (19.5 – 31.0 lbf ft). Lock the nut in position with a new split-pin.

11    If in doubt as to the fitted position of any wheel spacers, dust seals, etc, then refer to the appropriate figure accompanying this text. Finally, before taking the machine on the road, carry out a check to ensure that the wheel spins freely. Operate the brake several times to ensure its correct operation and recheck all disturbed connections for security.

3.6a Do not omit to refit the wheel spacer before fitting the wheel

3.6b Check the position of the speedometer gearbox ...

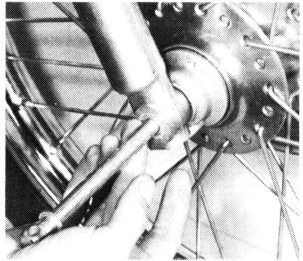

3.10a ... before inserting the wheel spindle with its washer

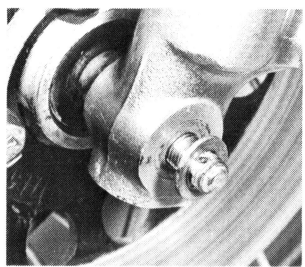

3.10b Fit the plain washer over the spindle ...

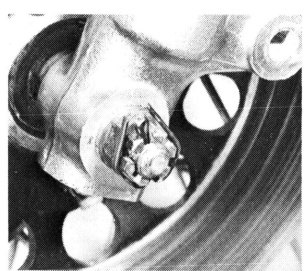

3.10c ... and lock the tightened nut in position with a new split-pin

3.10d Reconnect the speedometer drive cable

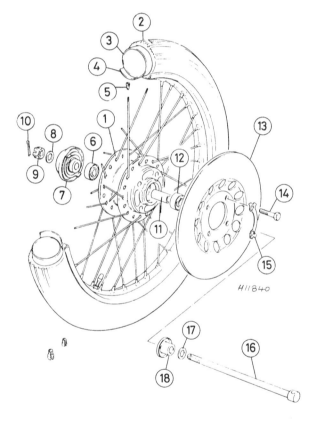

**Fig. 5.1 Front wheel – disc brake models**

| | | | |
|---|---|---|---|
| 1 | Hub | 10 | Split pin |
| 2 | Tyre | 11 | Spacer |
| 3 | Inner tube | 12 | Left-hand bearing |
| 4 | Rim tape | 13 | Brake disc |
| 5 | Spoke nipple | 14 | Bolt – 4 off |
| 6 | Right-hand bearing | 15 | Tab washer – 2 off |
| 7 | Speedometer gearbox | 16 | Spindle |
| 8 | Washer | 17 | Washer |
| 9 | Castellated nut | 18 | Shouldered spacer |

### 4 Speedometer drive gear: examination and renovation

**Disc brake models**

1 The speedometer gearbox fitted to machines equipped with a disc brake on the front wheel is a separate unit which locates over the wheel spindle, on the right-hand side of the wheel hub. Full details of removal and fitting of this unit are contained in Section 3 of this Chapter.

2 The nut can be partially dismantled for servicing by removing the circlip which retains the drive plate and its washer in position. Separate the washer and drive plate from the main body of the unit and wipe each component part clean.

3 Check the condition of the drive plate tangs; broken or badly worn tangs are the usual cause of the gearbox failing to drive the speedometer cable. The rubber seal within the unit body should show no signs of damage or deterioration. If any one part of the speedometer gearbox has failed, then the complete assembly must be renewed as Suzuki do not supply individual component parts for this unit. Before refitting the drive plate, pack the space between the rubber seal and the plate with a good quality high melting-point grease.

4.3c ... and locking the plate in position with the circlip

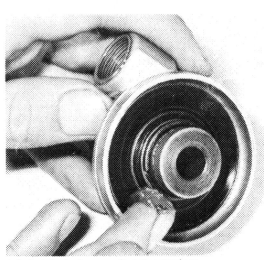

4.3a Lubricate the speedometer gearbox ...

4.3b ... before fitting a serviceable drive plate ...

**Drum brake models**

4 The speedometer gearbox fitted to machines equipped with a drum brake on the front wheel forms part of the brake backplate assembly. Refer to Sections 3 and 14 of this Chapter for details of front wheel and brake backplate removal.

5 To dismantle this assembly, remove the circlip from the centre of the backplate and lift the washer, drive plate and drivegear from position. Note that some assemblies may have a second washer located beneath the drive gear. Invert the backplate and remove the small grub screw from the side of the worm gear housing. The complete worm gear assembly can now be pulled from its housing. Take care to retain the two small thrust washers. Clean the component parts and lay them out on a clean work surface, ready for inspection.

6 Damage to any one component part should be immediately obvious and that part should be placed to one side, ready for renewal. Pay particular attention to the condition of the tangs on the drive plate; broken or badly worn tangs are the usual cause of the gearbox failing to drive the speedometer cable. Note also the condition of the large oil seal which surrounds the drive plate. If this seal shows signs of damage or deterioration, then it must be renewed otherwise grease will work through from the drive assembly to contaminate the brake linings.

7 Reassemble the component parts in the reverse order of dismantling. Lightly lubricate the drive assembly with a good quality high melting point grease and refer to the figure accompanying this text if in doubt as to the fitted position of any one part.

### 5 Front wheel bearings: removal, examination and fitting

1 Although there are various types of wheel and front brake assemblies fitted to the machines covered in this Manual, the procedure for removal and fitting of the wheel bearings is similar for each type. Access to these bearings may be gained after removal of the wheel from the forks.

2 On machines equipped with a drum brake, withdraw the brake assembly from the wheel hub and remove the dust cover and spacer from the opposite side of the hub. On machines equipped with a disc brake, detach the speedometer gearbox from the wheel hub. In order to avoid damage occurring to the disc during bearing removal, it is advisable to remove the disc in accordance with the instructions given in Section 6 of this Chapter.

3   Position the wheel on a work surface with its hub well supported by wooden blocks so that enough clearance is left beneath the wheel to drive the bearing out. Ensure the blocks are placed as close to the bearing as possible, to lessen the risk of distortion occurring to the hub casting whilst the bearings are being removed or fitted.

4   Place the end of a long-handled drift against the upper face of the lower bearing and tap the bearing downwards out of the wheel hub. The spacer located between the two bearings may be moved sideways slightly in order to allow the drift to be positioned against the face of the bearing. Move the drift around the face of the bearing whilst drifting it out of position, so that the bearing leaves the hub squarely.

5   With the one bearing removed, the wheel may be lifted and the spacer withdrawn from the hub. Invert the wheel and remove the second bearing, using a similar procedure to that used for the first. On machines equipped with a drum brake, the oil seal which fits against the right-hand bearing will be driven out as the bearing is removed. This seal should be closely inspected for any indication of damage, hardening or perishing and renewal if necessary. It is advisable to renew this seal as a matter of course if the bearings are found to be defective.

6   Remove all the old grease from the hub and bearings, giving the latter a final wash in petrol. Check the bearings for signs of play or roughness when they are turned. If there is any doubt about the condition of a bearing, it should be renewed.

7   If the original bearings are to be refitted, then they should be repacked with the recommended grease before being fitted into the hub. New bearings must also be packed with the recommended grease. Ensure that the bearing recesses in the hub are clean and both bearings and recess mating surfaces lightly greased to aid fitting. Check the condition of the hub recesses for evidence of abnormal wear which may have been caused by the outer race of a bearing spinning. If evidence of this happening is found, and the bearing is a loose fit in the hub, then it is best to seek advice from a Suzuki service agent or a competent motorcycle engineer. Do not proceed with fitting a bearing that is a loose fit in the hub.

8   With the wheel hub and bearing thus prepared, proceed to fit the bearings and central spacer as follows. With the hub again well supported by the wooden blocks, drift the first of the two bearings into position. To do this, use a soft-faced hammer in conjunction with a socket or length of metal tube which has an overall diameter which is slightly less than that of the outer race of the bearing. Invert the wheel, insert the spacer and fit

the second bearing, using the same procedure as given for the first. Take great care to ensure that each of the bearings enters its housing correctly, that is, square to the housing, otherwise the housing surface may be broached. On machines equipped with a drum brake, fit the oil seal against the right-hand bearing by using a method similar to that used for fitting each bearing.

9   On completion of fitting the bearings, reassemble the hub components. Take care to refit each component in its original location. If in doubt as to the position of a component, then refer to the appropriate figure accompanying this text. Details of refitting the brake disc are given in Section 6 of this Chapter.

## 6   Front brake disc: examination, removal and refitting

1   The brake disc can be checked for wear and for warpage whilst the front wheel is still in the machine. Using a micrometer, measure the thickness of the disc at the point of greatest wear. If the measurement is much less than the recommended service limit of 3.0 mm (0.12 in), then the disc should be renewed. Check the warpage (run out) of the disc by setting up a suitable pointer close to the outer periphery of the disc and spinning the front wheel slowly. If the total warpage is more than 0.30 mm (0.012 in), the disc should be renewed. A warped disc, apart from reducing the braking efficiency, is likely to cause juddering during braking and will also cause the brake to bind when it is not in use.

2   The brake disc should also be checked for bad scoring on its contact area with the brake pads. If any of the above mentioned faults are found, then the disc should be removed from the wheel for renewal.

3   To detach the disc, first remove the wheel by following the procedure given in Section 3 of this Chapter. The disc is retained in position by four bolts, each one of which is screwed directly into the wheel hub. These bolts are locked in pairs by tab washers. Bend down the ears of these washers and remove the bolts. The disc can now be eased off the hub boss.

4   Fit the disc by reversing the removal procedure. Do not reuse the same tabs at the tab washer ends, renew the washers if necessary. Avoid placing any strain on the disc by tightening the four retaining bolts evenly and in a diagonal sequence.

5.4 The wheel bearings may be drifted from position

6.1 Check the brake disc for wear

6.4 Lock the disc retaining bolts by bending the tab washer ends

## 7  Front disc brake: examination of the hydraulic hose

1    An external brake hose is used as a means of transmitting hydraulic pressure to the caliper unit once the front brake lever is applied. The hose is of the flexible type and is routed through a clamp which is secured to the steering head lower yoke.

2    When the brake assembly is being overhauled, or at any time during a routine maintenance or cleaning procedure, check the condition of the hose for signs of leakage, damage, deterioration or scuffing against any cycle components. The union connections at either end of the hose must also be in good condition, with no stripped threads or damaged sealing washers. Do not tighten these union bolts over the recommended torque setting of 2.5 – 4.0 kgf m (18.0 – 29.0 lbf ft) as they are easily sheared if overtightened.

3    Suzuki recommend that brake hoses be renewed every two years in the interests of safety.

## 8  Front disc brake: component examination and renewal of the brake pads

1    Check the front brake master cylinder, hose and caliper unit for signs of leakage. Pay particular attention to the condition of the hose, which should be renewed without question if there are signs of cracking, splitting or other exterior damage. With the machine positioned on its centre stand on an area of flat ground and with the handlebars positioned so that the front wheel is pointing forward, check the hydraulic fluid level by referring to the upper and lower level lines visible on the exterior of the transparent reservoir body.

2    Replenish the reservoir after first removing the cap on the brake fluid reservoir and lifting out the diaphragm. The cap is a screw on/off fitting and is retained in position by a lock plate which can be removed by unscrewing the mirror stem from the master cylinder casting and the small crosshead screw from the reservoir body. The condition of the fluid can be checked at the same time. Checking the fluid level is one of the maintenance tasks which should **never be neglected.** If the fluid is below the lower level mark, brake fluid of the correct specification must be added. **Never** use engine oil or any fluid other than that recommended. Other fluids have unsatisfactory characteristics and will rapidly destroy the seals. The fluid level is unlikely to fall other than a small amount, unless leakage has occurred somewhere in the system. If a rapid change of level is noted, a careful check for leaks should be made before the machine is used again. It is also worth noting that Suzuki recommend that

the brake hoses should be renewed every two years in the interests of safety.

3    The brake pads should be inspected for wear. Each has a red groove, which marks the wear limit of the friction material. When this limit is reached, both pads must be renewed, even if only one has reached the wear mark. In normal use, both pads will wear at the same rate and therefore both must be renewed. The degree of wear present on each brake pad can be checked by viewing the pads from the front of the machine.

4    If the brake action becomes spongy, or if any part of the hydraulic system is dismantled (such as when a hose has been renewed) it is necessary to bleed the system in order to remove all traces of air. Follow the procedures given in Section 11 of this Chapter.

5    Before attempting renewal of the brake pads, thoroughly clean the area around the brake caliper. This will prevent any ingress of road dirt into the caliper assembly. To gain access to the pads for renewal, detach the caliper assembly from the fork leg by removing the two securing bolts, each with its plain washer and spring washer. Disconnection of the hydraulic hose is not required.

6    Remove the single screw and the convolute backing plate from the inner side of the caliper unit. The inner pad is now free and may be displaced towards the centre of the caliper and lifted out. The outer pad which abuts against the caliper piston is not retained positively and may be lifted out.

7    Fit the new pads and refit the caliper assembly to the fork leg by reversing the dismantling procedure, whilst noting the following points. Take care not to operate inadvertently the brake lever at any time whilst the caliper assembly is removed from the brake disc. This will operate and displace the caliper piston, thereby making reassembly considerably more difficult. Before fitting the new pads, push the caliper piston inwards slightly so that there is sufficient clearance between the brake pads to allow the caliper to fit over the disc. It is recommended that the outer periphery of the outer (piston) pad is lightly coated with disc brake assembly grease (silicone grease). Use the grease sparingly and ensure that grease **DOES NOT** come in contact with the friction surface of the pad.

8    Before refitting the caliper to fork leg securing bolts, check that the spring washer located beneath each bolt head has not become flattened. If this is the case, then the washer must be renewed as it will no longer serve its locking function. Tighten each of these bolts to a torque loading of 1.5 – 2.5 kgf m (11.0 – 18.0 lbf ft).

9    Finally, in the interests of safety, always check the function of the brakes; pump the brake lever several times to restore full braking power, before taking the machine on the road.

8.1 Check the fluid level in the reservoir

8.2a The reservoir cap is retained in position by a lock plate

8.2b Remove the lock plate ...

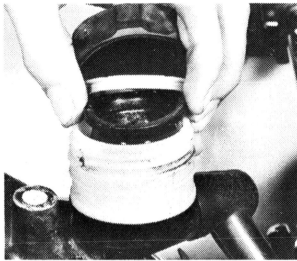

8.2c ... to allow removal of the cap and diaphragm assembly

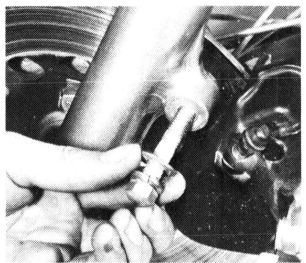

8.5 Each caliper securing bolt has a plain and a spring washer fitted to it

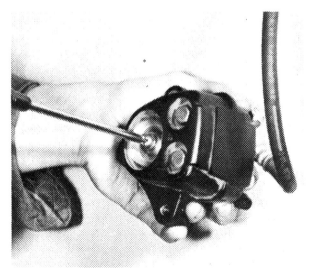

8.6a Remove the convolute backing plate ...

8.6b ... to facilitate removal of the inner brake pad

## 9 Front disc brake: dimantling, examination, renovation and reassembly of the caliper

1 Select a suitable receptacle into which to drain the brake fluid. Position the receptacle, together with some clean rag, beneath the caliper unit. Remove the bolt which retains the hose union to the caliper and allow the fluid to drain from the hose into the receptacle. Take great care not to allow the fluid to spill onto any paintwork; it is a very effective paint stripper. Hydraulic fluid will also damage rubber and plastic components. In the event of spillage, wipe up the fluid immediately by using the rag mentioned above.

2 Remove the caliper assembly from the fork leg and displace the brake pads by following the procedure given in the preceding Section.

3 Remove the two spindle bolts which pass through the caliper body and separate the body from the caliper support bracket. Using the flat of a small screwdriver, prise out the piston boot whilst taking great care not to scratch the surface of the cylinder bore.

4 The caliper piston can be displaced most easily by applying an air jet to the hydraulic fluid feed orifice. Take care not to use too high an air pressure whilst doing this and place a thick wad of clean rag over the end of the caliper bore; this will serve to both catch the piston and prevent any injury through hydraulic fluid being ejected from the caliper under pressure. It is still, however, advisable to wear some form of eye protection during this operation.

5 Clean the caliper components thoroughly, only in hydraulic brake fluid. **Never** use petrol or cleaning solvent for cleaning hydraulic brake parts otherwise the rubber components will be damaged. Discard all the rubber components as a matter of course. The replacement cost is relatively small and does not warrant re-use of components vital to safety. Check the piston and caliper cylinder bore for scoring, rusting or pitting. If any of these defects are evident it is unlikely that a good fluid seal can be maintained and for this reason the components should be renewed.

6 Inspect the shank of each spindle bolt for any signs of damage or corrosion and clean or renew each one as necessary. Remove the bleed screw and check that it has not become blocked. Check the condition of the screw sealing cap and renew it, if necessary.

7 Assemble the caliper unit by reversing the dismantling sequence. Note that assembly must be undertaken under ultra-clean conditions. Particles of dirt will score the bearing surfaces of moving parts and cause early failure.

8 When assembling the unit, pay attention to the following points. When fitting the new piston seal, take care to ensure that it is not twisted on its retaining groove. Apply a generous amount of new brake fluid to the surface of the caliper bore and to the periphery of the piston before pushing the piston slowly into position whilst taking care not to damage the piston seal.

9 Fit the two new O-rings to the larger of the two spindle bolts and then lubricate the shank of each bolt with Suzuki caliper caliper axle grease (part No 99000-25110) or a suitable equivalent brake assembly grease. This grease has high heat resistant qualities. Tighten each one of these bolts to a torque loading of 1.5 – 2.0 kgf m (11.0 – 14.5 lbf ft).

10 Refit the brake pads and refit the caliper assembly to the fork leg by following the procedure given in the preceding Section. Before reconnecting the brake hose union to the caliper, check the condition of the two gasket washers located one either side of the union. Renew these gasket washers if necessary and then fit and tighten the union bolt to a torque loading of 2.5 – 4.0 kgf m (18.0 – 29.0 lbf ft).

11 Refill the master cylinder reservoir with new hydraulic brake fluid and bleed the system by following the procedure given in Section 11 of this Chapter. On completion of bleeding, carry out a check for leakage of fluid whilst applying the brake lever. Push the machine forward and bring it to a halt by applying the brake. Do this several times to ensure that the brake is operating correctly before taking the machine for a test run. During this run, use the brakes as often as possible and on completion, recheck for signs of fluid loss.

12 The component parts of the caliper assembly (and the master cylinder assembly) may wear or deteriorate in function over a long period of use. It is however, generally difficult to foresee how long each component will work with proper efficiency. From a safety point of view it is best to change all the expendable parts every two years on a machine that has covered a normal mileage.

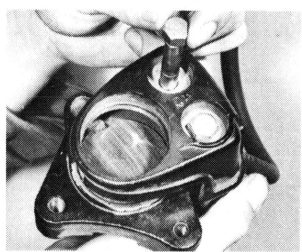

9.3 Remove the two caliper spindle bolts

## 10 Front disc brake: removal, examination, renovation and refitting of the master cylinder

1 The master cylinder and hydraulic reservoir take the form of a combined unit mounted on the right-hand side of the handlebars, to which the front brake lever is attached. The master cylinder is actuated by the front brake lever, and applies hydraulic pressure through the system to operate the front brake when the handlebar lever is manipulated. The master cylinder pressurises the hydraulic fluid in the brake pipe which, being incompressible, causes the piston to move in the caliper unit and apply the friction pads to the brake disc. If the master cylinder seals leak, hydraulic pressure will be lost and the braking action rendered much less effective.

2 Before the master cylinder can be removed, the system must be drained. Place a clean container below the caliper unit and attach a plastic tube from the bleed screw on top of the caliper unit to the container. Open the bleed screw one complete turn and drain the system by operating the brake lever until the master cylinder reservoir is empty. Close the bleed screw and remove the pipe.

3 Position a pad of clean rag beneath the point where the brake hose joins the master cylinder. This simple precaution is essential to prevent brake fluid from dripping onto, and there-fore damaging, any plastic and painted components located beneath the hose union once the union bolt is removed. Detach the rubber cover from the head of the union bolt and remove the bolt. Once any excess fluid has drained from the union connection, wrap the end of the hose in rag or polythene and then attach it to a point on the handlebars.

4 Unscrew the rear view mirror stem from the body of the

master cylinder and remove the lock plate by unscrewing its single retaining screw from the reservoir body. Detach the front brake lever from its pivot point on the master cylinder body by removing its retaining bolt and nut.

5    Support the unit body so that it remains upright and remove the two bolts that secure it to the handlebar. Carefully lift the unit clear of the machine, remove the reservoir cap, diaphragm and gasket and empty any surplus fluid left within the unit into the container used during the draining procedure. Remove the piston components, starting with the boot and the circlip. Use the flat of a small screwdriver to carefully ease the boot out of position and a pair of straight-nose circlip pliers to remove the circlip. Withdraw the piston, the primary cup and the spring.

6    Detach the reservoir from the unit body by first unscrewing the two crosshead retaining screws from its base and then easing it out of its recess. Where the reservoir has a retaining plate fitted beneath the two screws, invert the reservoir and tap it into a cupped hand so that the plate falls from position. Note that this plate has a third hole drilled through its edge; this hole must align with the corresponding hole in the unit body. Using the flat of a small screwdriver, carefully remove the O-ring from its recess in the unit body.

7    Place all the master cylinder component parts in a clean container and wash each part thoroughly in new brake fluid. Lay the parts out on a sheet of clean paper and examine each one as follows.

8    Inspect the unit body for signs of stress failure around both the brake lever pivot lugs and the handlebar mounting points. Carry out a similar inspection around the hose union boss. Examine the cylinder bore for signs of scoring or pitting. If any of these faults are found, then the unit body must be renewed.

9    Inspect the surface of the piston for signs of scoring or pitting and renew it if necessary. It is advisable to discard all of the rubber components of the piston assembly as a matter of course as the replacement cost is relatively small and does not warrant re-use of components vital to safety. The same advice applies to the O-ring fitted between the reservoir and the unit body. Inspect the threads of the brake hose union bolt for any signs of failure and renew the bolt if in the slightest doubt. Renew each of the gasket washers located one either side of the hose union.

10  Check before reassembly that any traces of contamination remaining within the reservoir body have been removed. Inspect the diaphragm to see that it is not perished or split. It must be noted at this point that any reassembly work must be undertaken in ultra-clean conditions. Particles of dirt entering the component will only serve to score the working points of the cylinder and thereby cause early failure of the system.

11  When reassembling and fitting the master cylinder, follow the removal and dismantling procedures in the reverse order whilst paying particular to the following points. Make sure that the piston components are fitted the correct way round and in the correct order. Immerse all of these components in new brake fluid prior to reassembly and refer to the figure accompanying this text when in doubt as to their fitted positions. Tighten the two bolts that retain the unit to the handlebars to a torque loading of 0.6 – 0.9 kgf m (4.5 – 6.5 lbf ft) and the hose union bolt to a torque loading of 2.5 – 4.0 kgf m (18.0 – 29.0 lbf ft).

12  Bleed the brake system after refilling the reservoir with new hydraulic fluid, then check for leakage of fluid whilst applying the brake lever. Push the machine forward and bring it to a halt by applying the brake. Do this several times to ensure that the brake is operating correctly before taking the machine for a test run. During the run, use the brakes as often as possible and on completion, recheck for signs of fluid loss.

13  The component parts of the master cylinder assembly (and the caliper assembly) may wear or deteriorate in function over a long period of use. It is however, generally difficult to foresee how long each component will work with proper efficiency and from a safety point of view, it is best to change all the expendable parts every two years on a machine that has covered a normal mileage.

## 11  Front disc brake: bleeding the hydraulic system

1    As mentioned earlier, brake action is impaired or even rendered inoperative if air is introduced into the hydraulic system. This can occur if the seals leak, the reservoir is allowed to run dry or if the system is drained prior to the dismantling of any component part of the system. Even when the system is refilled with hydraulic fluid, air pockets will remain and because air will compress, the hydraulic action is lost.

2    Check the fluid content of the reservoir and fill almost to the top. Remember that hydraulic brake fluid is an excellent paint stripper, so beware of spillage, especially near the fuel tank. Do not omit to refit the reservoir cap to prevent dirt from entering the system.

3    Place a clean glass jar below the brake caliper unit and attach a clear plastic tube from the caliper bleed screw to the container. Place some clean hydraulic fluid in the container so that the pipe is always immersed below the surface of the fluid.

4    Pump the brake lever several times in rapid succession, finally holding it in the 'fully on' position. Loosen the bleed screw one half of a turn so that the brake fluid is seen to run down the tube into the container. This will cause the pressure within the system to be released, thereby causing the brake lever to move so that it touches the throttle twistgrip. Directly this happens, nip the bleed screw tight and then release the lever. As the fluid is ejected from the bleed screw the level in the reservoir will fall. Take care that the level does not drop too low whilst the operation continues, otherwise air will re-enter the system necessitating a fresh start.

5    Repeat the above procedure until no further air bubbles emerge from the end of the plastic pipe. Hold the brake lever against the twistgrip and tighten the caliper bleed screw. Remove the plastic tube after the bleed screw is closed.

6    Check the brake action for sponginess, which usually denotes there is still air in the system. If the action is spongy, continue the bleeding operation in the same manner, until all traces of air are removed.

7    Refit the sealing cap to the bleed screw. Bring the fluid in the reservoir up to the correct level and refit the diaphragm, sealing gasket and cap. Check the complete system for leaks and recheck the brake action.

8    Note that fluid from the container placed below the brake caliper unit whilst the system is bled, should not be reused, as it will have become aerated and may have absorbed moisture.

11.3 Bleeding the front brake hydraulic system

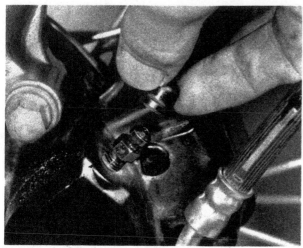

11.7 Do not omit to refit the bleed screw sealing cap

13.1 An indication of brake shoe lining wear is provided

## 12 Front drum brake: adjustment

1   Adjustment of the front drum brake is correct when there is 20 – 30 mm (0.8 – 1.2 in) of clearance between the end of the handlebar lever and the throttle twistgrip with the lever fully applied.

2   To adjust the clearance between the lever and twistgrip, simply turn the nut at the cam shaft operating arm end of the cable the required amount in the appropriate direction. Any minor adjustments necessary may then be made with the cable adjuster at the handlebar lever bracket. To use this adjuster, simply loosen the lock ring, turn the knurled adjuster the required amount and then retighten the lock ring.

3   On completion of adjustment, check the brake for correct operation by spinning the wheel and applying the brake lever. There should be no indication of the brake binding as the wheel is spun. If the brake shoes are heard to be brushing against the surface of the wheel drum back off on the cable adjuster slightly until all indication of binding disappears. The brake may be readjusted after a period of bedding-in has been allowed for the brake shoes.

## 13 Front and rear drum brakes: checking shoe wear

1   An indication of brake shoe lining wear is provided by an indicator line which is cast into the brake backplate. If, when the brake is correctly aligned and applied fully, the line on the end of the brake cam spindle is seen to align with a point outside the arc shown by the indicator line, then the lining on the brake shoes can be assumed to have worn beyond limits and should be renewed at the earliest possible opportunity.

## 14 Front and rear drum brake assemblies: dismantling, examination, renovation and reassembly

1   The brake assembly, complete with the brake backplate, can be withdrawn from its wheel hub after removal of the wheel from the machine. With the wheel laid on a work surface, brake backplate uppermost, the brake backplate may be lifted away from the hub. It will come away quite easily, with the brake shoe assembly attached to its back.

2   Examine the condition of the brake linings. If they are worn beyond the specified limit then the brake shoes should be renewed. The linings are bonded on and cannot be supplied separately.

3   If oil or grease from the wheel bearings has badly contaminated the linings, the brake shoes should be renewed. There is no satisfactory way of degreasing the lining material. Any surface dirt on the linings can be removed with a stiff-bristled brush. High spots on the linings should be carefully eased down with emery cloth.

4   Examine the drum surface for signs of scoring, wear beyond the service limit or oil contamination. All of these conditions will impair braking efficiency. Remove all traces of dust, preferably using a brass wire brush, taking care not to inhale any of it, as it is of an asbestos nature, and consequently harmful. Remove oil or grease deposits, using a petrol soaked rag.

5   If deep scoring is evident, due to the linings having worn through to the shoe at some time, the drum must be skimmed on a lathe, or renewed. Whilst there are firms who will undertake to skim a drum whilst it is fitted to the wheel, it should be borne in mind that excessive skimming will change the radius of the drum in relation to the brake shoes, therefore reducing the friction area until extensive bedding in has taken place. Also full adjustment of the shoes may not be possible. If in doubt about this point, the advice of one of the specialist engineering firms who undertake this work should be sought.

6   Note that it is a false economy to try to cut corners with brake components; the whole safety of both machine and rider being dependent on their good condition.

7   Removal of the brakes shoes is accomplished by folding the shoes together so that they form a 'V'. With the spring tension relaxed, both shoes and springs may be removed from the brake backplates as an assembly. Detach the springs from the shoes and carefully inspect them for any signs of fatigue or failure. If in doubt, compare them with a new set of springs.

8   Before fitting the brake shoes, check that the brake operating cam is working smoothly and is not binding in its pivot. The cam can be removed by withdrawing the retaining bolt on the operating arm and pulling the arm off the shaft. Before removing the arm, it is advisable to mark its position in relation to the shaft, so that it can be relocated correctly.

9   Remove any deposits of hardened grease or corrosion from the bearing surface of the brake cam shoe by rubbing it lightly with a strip of fine emery paper or by applying solvent with a piece of rag. Lightly grease the length of the shaft and the face of the operating cam prior to reassembly. Clean and grease the pivot stub which is set in the backplate.

10  Check the condition of the O-ring which prevents the escape of grease from the end of the cam shaft. If it is in any way damaged or perished, then it must be renewed before the

shaft is relocated in the backplate. Relocate the cam shaft and align and fit the operating arm with the O-ring and plain washer. The bolt and nut retaining the arm in position on the shaft should be torque loaded to 0.5 – 0.8 kgf m (3.5 – 6.0 lbf ft).

11  Before refitting existing shoes, roughen the lining surface sufficiently to break the glaze which will have formed in use. Glasspaper or emery cloth is ideal for this purpose but take care not to inhale any of the asbestos dust that may come from the lining surface.

12  Fitting the brake shoes and springs to the brake backplate is a reversal of the removal procedure. Some patience will be needed to align the assembly with the pivot and operating cam whilst still retaining the springs in position; once they are correctly aligned though, they can be pushed back into position by pressing downwards in order to snap them into position. Do not use excessive force, or there is risk of distorting the brake shoes permanently.

14.1 The brake assembly can be withdrawn from the wheel hub

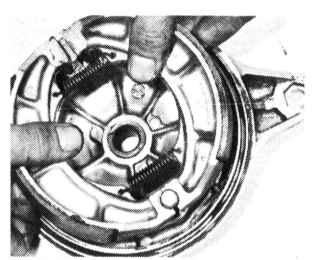

14.7 Fold the brake shoes together to aid removal

14.9a Lightly grease the pivot stub ...

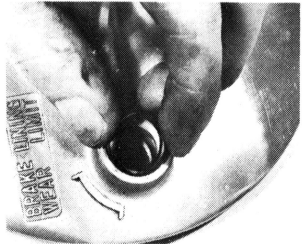

14.9b ... and fit a serviceable O-ring ...

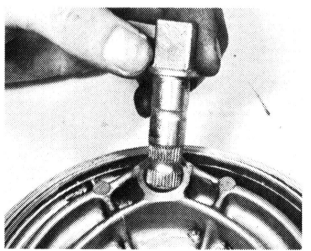

14.9c ... before inserting the greased brake cam shaft into the backplate

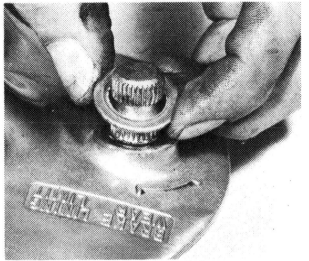

14.10a Refit the plain washer over the O-ring ...

14.10b ... and secure the operating arm in position

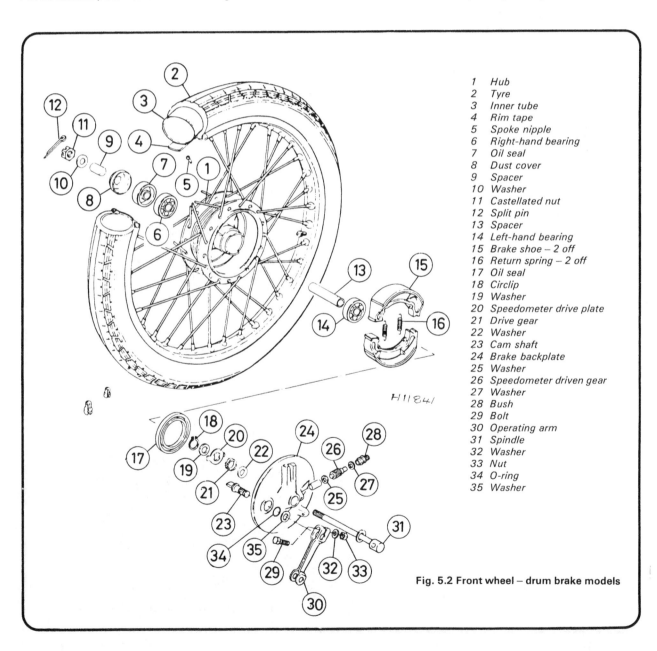

| 1 | Hub |
|---|---|
| 2 | Tyre |
| 3 | Inner tube |
| 4 | Rim tape |
| 5 | Spoke nipple |
| 6 | Right-hand bearing |
| 7 | Oil seal |
| 8 | Dust cover |
| 9 | Spacer |
| 10 | Washer |
| 11 | Castellated nut |
| 12 | Split pin |
| 13 | Spacer |
| 14 | Left-hand bearing |
| 15 | Brake shoe – 2 off |
| 16 | Return spring – 2 off |
| 17 | Oil seal |
| 18 | Circlip |
| 19 | Washer |
| 20 | Speedometer drive plate |
| 21 | Drive gear |
| 22 | Washer |
| 23 | Cam shaft |
| 24 | Brake backplate |
| 25 | Washer |
| 26 | Speedometer driven gear |
| 27 | Washer |
| 28 | Bush |
| 29 | Bolt |
| 30 | Operating arm |
| 31 | Spindle |
| 32 | Washer |
| 33 | Nut |
| 34 | O-ring |
| 35 | Washer |

H11841

**Fig. 5.2 Front wheel – drum brake models**

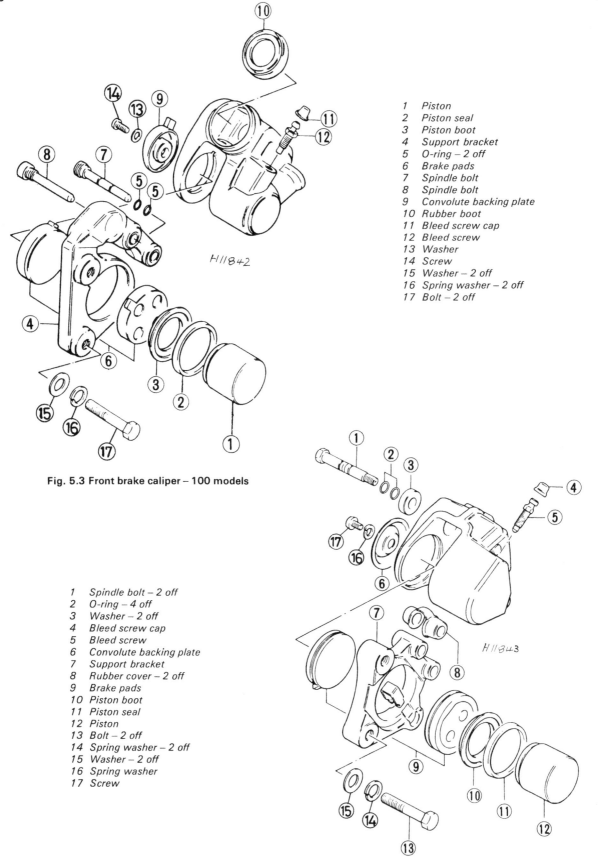

1   Piston
2   Piston seal
3   Piston boot
4   Support bracket
5   O-ring – 2 off
6   Brake pads
7   Spindle bolt
8   Spindle bolt
9   Convolute backing plate
10  Rubber boot
11  Bleed screw cap
12  Bleed screw
13  Washer
14  Screw
15  Washer – 2 off
16  Spring washer – 2 off
17  Bolt – 2 off

H11842

**Fig. 5.3 Front brake caliper – 100 models**

1   Spindle bolt – 2 off
2   O-ring – 4 off
3   Washer – 2 off
4   Bleed screw cap
5   Bleed screw
6   Convolute backing plate
7   Support bracket
8   Rubber cover – 2 off
9   Brake pads
10  Piston boot
11  Piston seal
12  Piston
13  Bolt – 2 off
14  Spring washer – 2 off
15  Washer – 2 off
16  Spring washer
17  Screw

H11843

**Fig. 5.4 Front brake caliper – 125 models**

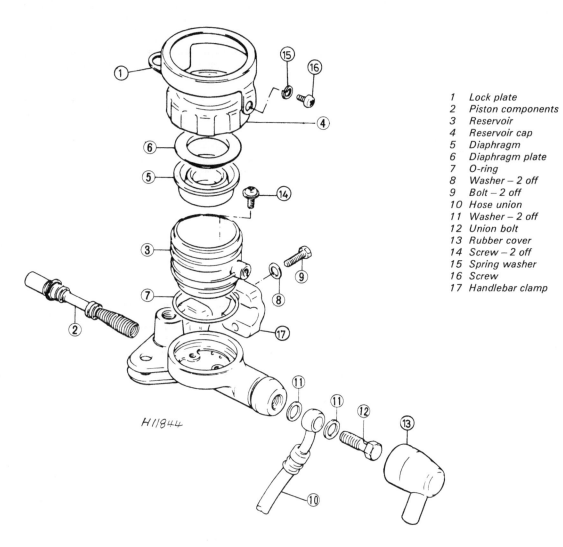

1 Lock plate
2 Piston components
3 Reservoir
4 Reservoir cap
5 Diaphragm
6 Diaphragm plate
7 O-ring
8 Washer – 2 off
9 Bolt – 2 off
10 Hose union
11 Washer – 2 off
12 Union bolt
13 Rubber cover
14 Screw – 2 off
15 Spring washer
16 Screw
17 Handlebar clamp

H11844

Fig. 5.5 Front brake master cylinder

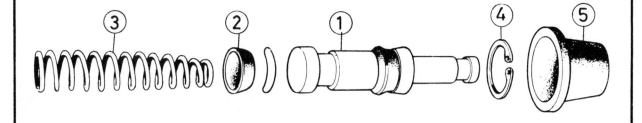

Fig. 5.6 Master cylinder piston components arrangement

1 Piston
2 Primary cup
3 Spring
4 Circlip
5 Boot

## 15 Rear brake: adjustment

1    Adjustment of the rear brake is correct when there is 20 – 30 mm (0.8 – 1.2 in) of movement, measured at the forward point of the brake pedal, between the point at which the brake pedal is fully depressed and the point where it abuts against its return stop.
2    The range of pedal movement may be adjusted simply by turning the nut on the wheel end of the brake operating rod in the required direction.
3    On completion of brake adjustment, check that the stop lamp switch operates the stop lamp as soon as the brake pedal is depressed. If necessary, adjust the height setting of the switch in accordance with the instructions given in Chapter 6.

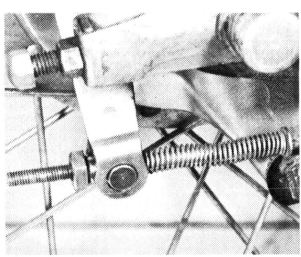

15.2 Adjust the rear brake by turning the brake rod nut

## 16 Rear wheel : examination, removal, renovation and fitting

1    Place the machine on its centre stand so that the rear wheel is raised clear of the ground. Check the condition of the wheel as described in Section 2 of this Chapter.
2    Before attempting wheel removal, it is first necessary to gain access to the right-hand side of the wheel spindle. This involves removal of the complete exhaust system from the machine. Full details of removing and fitting the exhaust system are given in Section 17 of Chapter 2.
3    With the exhaust system thus removed, insert a tommy bar through the hole provided in the end of the wheel spindle, displace the split-pin from the spindle retaining nut and remove the nut together with any plain washers located beneath it. Detach the torque arm from the brake backplate by removing the split-pin which passes through its retaining bolt and then removing the retaining nut and spring washer to allow the bolt to be withdrawn. Detach the brake operating rod from the cam shaft operating arm by removing the nut from the threaded end of the rod and then depressing the brake pedal so that the rod leaves the trunnion in the arm. Refit the nut, the trunnion from the operating arm, the return spring and the washer to the rod to prevent loss and then rest the end of the rod on the ground beneath the machine.
4    Place a piece of clean rag or paper beneath the forward section of swinging arm and, using the tommy bar as a handle, pull the wheel spindle clear of the machine. Support the wheel as it drops clear of the swinging arm fork legs. Remove the

spacer from the right-hand side of the wheel hub and ease the wheel over to the right until it abuts against the leg of the swinging arm fork. The rear sprocket, together with the cush drive hub and the final drive chain, may now be detached from the wheel hub and placed on the piece of rag or paper placed beneath the swinging arm. Take care to avoid any ingress of dirt into this assembly. Tilt the wheel slightly and ease it clear of the machine.
5    Once the faults found in the wheel have been rectified, the wheel may be refitted by reversing the procedure used for removal. Take note of the information given in Section 18 of this Chapter concerning the cush drive assembly and ensure that the wheel spacers and washers are correctly positioned before inserting the wheel spindle. If in doubt as to the fitted position of any components, refer to the figure accompanying this text.
6    When reconnecting the brake operating rod, feed the rod through the brake lever trunnion and fit the adjusting nut onto the end of the rod to retain it in position. Ensure that the spring is correctly positioned between the washer on the operating rod and the forward facing face of the lever trunnion.
7    Tighten the torque arm retaining nut to the specified setting of 1.0 – 1.5 kgf m (7.0 – 11.0 lbf ft) and fit a new split-pin through the hole in the shank of the bolt. Note that the torque arm must be properly attached to the brake backplate. Failure to ensure this will mean that on the first application of the rear brake, the wheel will lock, with disastrous consequences.
8    Give the end of the wheel spindle a light tap with a soft-faced hammer to seat it and refit the spindle retaining nut together with any plain washers. Tighten the nut to a torque setting of 3.6 – 5.2 kgf m (26.0 – 37.5 lbf ft) after having checked the tension of the final drive chain in accordance with the instructions given in Section 20 of this Chapter. Lock the nut in position with a new split pin, refit the exhaust system, carry out a check to ensure that the wheel spins freely and then check that the rear brake is correctly adjusted by following the instructions given in Section 15 of this Chapter. Before taking the machine on the road, operate the brake several times and recheck all disturbed connections for security.

## 17 Rear wheel bearing : removal, examination and fitting

1    The rear wheel assembly has three journal ball bearings. One bearing lies each side of the wheel hub and the third bearing is fitted in the cush drive assembly to which is attached the sprocket. Access to these bearings may be gained after removal of the wheel from the swinging arm fork and detachment of the cush drive assembly from the wheel hub.
2    The two bearings contained within the wheel hub may be drifted out of position by using a procedure similar to that given for removal of the front wheel bearings. It will, of course, be necessary to detach the brake backplate assembly from the wheel hub before commencing this operation.
3    Before attempting removal of the cush drive bearing, remove the stepped hollow spacer from its centre. Suzuki recommend that the sprocket should be removed from the cush drive hub but in practice it was found that this was not necessary, provided that care was taken in supporting the assembly during removal of the bearing. The oil seal contained in the cush drive hub will be drifted out of position along with the bearing. It is good practice to renew this seal whenever the bearing is found to be defective.
4    The procedure for examination and fitting of the bearings should be adopted from the relevant paragraphs in Section 5 of this Chapter. Make reference to the figure accompanying this text when refitting the hub component parts and take note of the information given in Section 18 of this Chapter concerning the cush drive assembly. Note that the two bearings fitted in the wheel hub are of different types and should not be exchanged with each other. Finally, do not omit to apply a light smear of grease to the tip of the cush drive hub oil seal before refitting the wheel to the machine.

16.3a Hold the wheel spindle in position ...

16.3b ... and remove the spindle retaining nut, together with its washers

16.5 Check the fitted position of the wheel spacer and spindle washers

16.7 Tighten the torque arm retaining nut

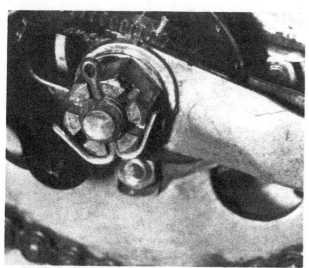

16.8 Lock the spindle retaining nut with a new split-pin

17.3a Drift the cush drive bearing from position

17.3b If necessary, the bearing seal can be levered from position

17.4a Do not omit to refit the central spacer

17.4b Lubricate each wheel bearing ...

17.4c ... before drifting the bearing into the hub centre

17.4d Refit the stepped hollow spacer ...

17.4e ... before inserting the cush drive hub into position

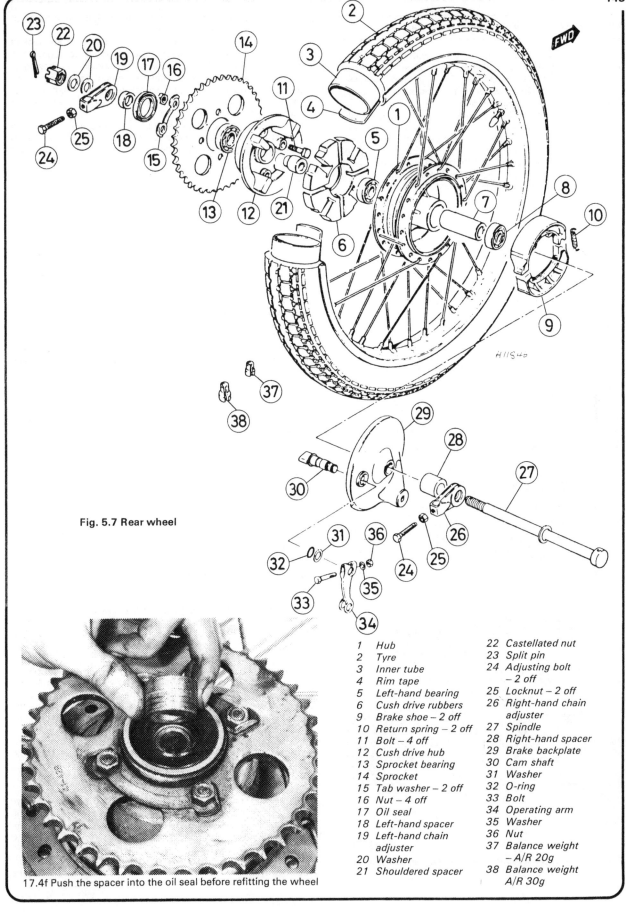

Fig. 5.7 Rear wheel

17.4f Push the spacer into the oil seal before refitting the wheel

1   Hub
2   Tyre
3   Inner tube
4   Rim tape
5   Left-hand bearing
6   Cush drive rubbers
9   Brake shoe – 2 off
10  Return spring – 2 off
11  Bolt – 4 off
12  Cush drive hub
13  Sprocket bearing
14  Sprocket
15  Tab washer – 2 off
16  Nut – 4 off
17  Oil seal
18  Left-hand spacer
19  Left-hand chain
    adjuster
20  Washer
21  Shouldered spacer

22  Castellated nut
23  Split pin
24  Adjusting bolt
    – 2 off
25  Locknut – 2 off
26  Right-hand chain
    adjuster
27  Spindle
28  Right-hand spacer
29  Brake backplate
30  Cam shaft
31  Washer
32  O-ring
33  Bolt
34  Operating arm
35  Washer
36  Nut
37  Balance weight
    – A/R 20g
38  Balance weight
    A/R 30g

## 18 Cush drive : examination and renovation

1   The cush drive assembly is contained within the left-hand side of the rear wheel hub. It takes the form of four rubber pads which are a push fit in the wheel hub. The cush drive hub, which is bolted to the rear wheel sprocket, incorporates four substantial vanes each one of which engages with the central slot in one of the rubber pads. This arrangement permits the sprocket to move within certain limits, therefore absorbing any surge or roughness in the transmission. The rubbers should be renewed when movement of the sprocket in relation to the wheel indicates bad compaction of the rubbers. The rubbers should also be renewed if they are seen to be breaking up.
2   It should be noted that it may be difficult to insert the vanes of the cush drive into new rubbers unless the rubbers are first lubricated with a solution of soapy water around the area of the slots.

## 19 Rear wheel sprocket : examination and renewal

1   The rear wheel sprocket is secured to the cush drive hub by four bolts whose nuts are locked in pairs by tab washers. To

18.1 The cush drive assembly incorporates four rubber pads

19.3 Lock the sprocket retaining nuts in position by bending up the tab washer ends

remove the sprocket from the hub, bend back the locking tab at each end of the washers and remove the nuts.
2   The sprocket need only be renewed if the teeth are hooked or badly worn. It is considered bad practice to renew one sprocket on its own; both drive sprockets should be renewed as a pair, preferably with a new final drive chain. If this recommendation is not observed, rapid wear resulting from the running of old and new parts together will necessitate even earlier replacement on the next occasion.
3   Fitting of the sprocket to the cush drive hub is simply a reversal of the removal procedure. Do not rebend the locking tab previously used; if necessary, renew the tab washers as a set.

## 20 Final drive chain : examination, adjustment and lubrication

1   As the final drive chain is fully exposed on all models it requires lubrication and adjustment at regular intervals. To adjust the chain, place the machine on its centre stand, take out the split-pin from the rear wheel spindle and slacken the spindle nut. Slacken also the nuts securing the brake torque arm. Undo the locknut on the chain adjusters and turn the adjuster bolts inwards to tighten the chain. Marks on the adjusters must be in line with identical marks on the frame fork to align the rear wheel correctly. A final check can be made by laying a straight wooden plank alongside the wheels, each side in turn. Chain tension is correct if there is 15 – 20 mm (0.6 – 0.8 in) of slack in the centre of the lever chain run.
2   Under no circumstances run the chain overtight to compensate for uneven wear. A tight chain will place excessive stresses on the gearbox and rear wheel bearings leading to their early failure. It will also absorb a surprising amount of power.
3   The chain may be checked for wear with it fitted to the machine. Commence by washing the chain thoroughly with a petrol-soaked brush. Wipe the chain dry and stretch it to its full length by turning the adjuster bolts inwards. Select a length of chain in the middle of the lower run and count of 21 pins, that is, a 20 pitch length. Measure the distance between the 1st and 21st pins. If this distance is found to be in excess of 259 mm (10.2 in) at any point along the chain's length, then the chain must be renewed.
4   Should the chain require renewal, move the rear wheel fully forward and then rotate it until the split link appears at the rear sprocket. Using a pair of flat-nosed pliers, remove the spring clip from the link and withdraw the link from the chain. The new chain should now be attached to the end of the upper run of old chain and the old chain used to pull the new chain into position over the gearbox sprocket. Once in position, connect the ends of the new chain with the new split link provided. Note that the spring clip which retains the link in position must have the side plate fitted beneath it, be seated correctly and have its closed end facing the direction of chain travel.
5   Note that replacement chains are now available in standard metric sizes from Renold Limited, the British chain manufacturer. When ordering a new chain, always quote the size, the number of chain links and the type of machine to which the chain is to be fitted.
6   Remember, on completion of examination and if necessary, adjustment or replacement of the chain, to ensure that both adjuster bolts are locked in position by tightening their locknuts. The wheel spindle retaining nut should be tightened to a torque setting of 3.6 – 5.2 kgf m (26.0 – 37.5 lbf ft) and locked in position by fitting a new split-pin. Finally, spin the rear wheel to ensure that it rotates freely, adjust the rear brake operating mechanism, when necessary, and retighten the torque arm retaining nuts to a torque setting of 1.0 – 1.5 kgf m (7.0 to 11.0 lbf ft).
7   After a period of running, the chain will require lubrication. Lack of oil will accelerate the rate of wear of both chain and

sprockets and will lead to harsh transmission. The application of engine oil will act as a temporary expedient, but it is preferable to use one of the proprietary graphited greases contained within an aerosol can. This type of lubricant is thrown off the chain less easily than engine oil. Ideally the chain should be removed at regular intervals, and immersed in a molten lubricant such as Linklyfe or Chainguard after it has been cleaned in a paraffin bath. These latter lubricants achieve better penetration of the chain links and rollers and are less likely to be thrown off when the chain is in motion.

20.6 Lock each adjuster bolt in position by tightening its locknut

20.7 An aerosol chain lubricant should be used at frequent intervals

## 21 Tyres : removal and fitting

1    At some time or other the need will arise to remove and replace the tyres, either as a result of a puncture or because replacements are necessary to offset wear. To the inexperienced, tyre changing represents a formidable task, yet if a few simple rules are observed and the technique learned the whole operation is surprisingly simple.

2    To remove the tyre from either wheel first detach the wheel from the machine. Deflate the tyre by removing the valve insert and when it is fully deflated, push the bead from the tyre away from the wheel rim on both sides so that the bead enters the centre well of the rim. Remove the locking cap and push the tyre valve into the tyre itself.

3    Insert a tyre lever close to the valve and lever the edge of the tyre over the outside of the wheel rim. Very little force should be necessary; if resistance is encountered it is probably due to the fact that the tyre beads have not entered the well of the wheel rim all the way round the tyre. Note that where the machine is fitted with cast alloy wheels, the risk of damage to the wheel rim can be minimised by the use of proprietary plastic rim protectors placed over the rim flange at the point where the tyre levers are inserted. Suitable rim protectors may be fabricated very easily from short lengths (4 - 6 inches) of thick-walled nylon petrol pipe which have been split down one side using a sharp knife. The use of rim protectors should be adopted whenever levers are used and, therefore, when the risk of damage is likely.

4    Once the tyre has been edged over the wheel rim, it is easy to work around the wheel rim so that the tyre is completely free to one side. At this stage, the inner tube can be removed.

5    Working from the other side of the wheel, ease the other edge of the tyre over the outside of the wheel rim that is furthest away. Continue to work around the rim until the tyre is free completely from the rim.

6    If a puncture has necessitated the removal of the tyre, reinflate the inner tube and immerse it in a bowl of water to trace the source of the leak. Mark its position and deflate the tube. Dry the tube and clean the area around the puncture with a petrol soaked rag. When the surface has dried, apply rubber solution and allow this to dry before removing the backing from the patch and applying the patch to the surface.

7    It is best to use a patch of self-vulcanising type, which will form a very permanent repair. Note that it may be necessary to remove a protective covering from the top surface of the patch, after it has sealed into position. Inner tubes made from synthetic rubber may require a special type of patch and adhesive, if a satisfactory bond is to be achieved.

8    Before refitting the tyre, check the inside to make sure that the object which caused the puncture is not trapped. Check the outside of the tyre, particularly the tread area, to make sure nothing is trapped that may cause a further puncture.

9    If the inner tube has been patched on a number of past occasions, or if there is a tear or large hole, it is preferable to discard it and fit a new one. Sudden deflation may cause an accident, particularly if it occurs with the front wheel.

10  To fit the tyre, inflate the inner tube sufficiently for it to assume a circular shape but only just. Then push it into the tyre so that it is enclosed completely. Lay the tyre on the wheel at an angle and insert the valve through the rim tape and the hole in the wheel rim. Attach the locking cap on the first few threads, sufficient to hold the valve captive in its correct location.

11  Starting at the point furthest from the valve, push the tyre bead over the edge of the wheel rim until it is located in the central well. Continue to work around the tyre in this fashion until the whole of one side of the tyre is on the rim. It may be necessary to use a tyre lever during the final stages.

12  Make sure that there is no pull on the tyre valve and again commencing with the area furthest from the valve, ease the other bead of the tyre over the edge of the rim. Finish with the area close to the valve, pushing the valve up into the tyre until the locking cap touches the rim. This will ensure the inner tube is not trapped when the last section of the bead is edged over the rim with a tyre lever.

13  Check that the inner tube is not trapped at any point. Reinflate the inner tube, and check that the tyre is sealing correctly around the wheel rim. There should be a thin rib moulded around the wall of the tyre on both sides, which should be equidistant from the wheel rim at all points. If the tyre is unevenly located on the rim, try bouncing the wheel when the tyre is at the recommended pressure. It is probable that one of the beads has not pulled clear of the centre well.

14  Always run the tyres at the recommended pressures and never under or over-inflate. The correct pressures are given in the Specifications Section of this Chapter.

15  Tyre replacement is aided by dusting the side walls particularly in the vicinity of the beads, with a liberal coating of french chalk. Washing-up liquid can also be used to good effect, but this has the disadvantage (where steel rims are used) of causing the inner surfaces of the wheel rim to corrode. Do not be over generous in the application of lubricant or tyre creep may occur.

16  On machines equipped with wire-spoked wheels, never fit the inner tube and tyre without the rim tape in position. If this procedure is overlooked there is good chance of the ends of the spoke nipples chafing the inner tube and causing a crop of punctures.

17  Never fit a tyre that has a damaged tread or side walls. Apart from the legal aspects, there is a very great risk of a blowout, which can have serious consequences on any two-wheel vehicle.

18  Tyre valves rarely give trouble, but it is always advisable to check whether the valve itself is leaking before removing the tyre. Do not forget to fit the dust cap, which forms an effective second seal.

21.14 Check the tyre pressures with an accurate gauge

## 22  Tyre valve dust caps

1  Tyre valve dust caps are often left off when a tyre has been replaced, despite the fact that they serve an important two-fold function. Firstly they prevent dirt or other foreign matter from entering the valve and causing the valve to stick open when the tyre pump is next applied. Secondly, they form an effective second seal so that in the event of the tyre valve sticking, air will not be lost.

2  Note that when a dust cap is fitted for the first time, the wheel may have to be rebalanced.

## 23  Wheel balancing

1  It is customary on all high perfomance machines to balance the front wheel complete with tyre and tube. The out of balance forces which exist are eliminated and the handling of the machine is improved in consequence. A wheel which is badly out of balance produces, through the steering, a most unpleasant hammering effect at high speeds.

2  Some tyres have a balance mark on the sidewall, usually in the form of a coloured dot. This mark must be in line with the tyre valve, when the tyre is fitted to the inner tube. Even then, the wheel may require the addition of balance weights, to offset the weight of the tyre valve itself.

3  If the front wheel is raised clear of the ground and is spun, it will come to rest with the tyre valve or the heaviest part downward and will always settle in the same position. Balance weights must be added to a point diametrically opposite this heavy spot until the wheel will come to rest in ANY position after it is spun.

4  For machines fitted with the conventional type of wire-spoked wheels, balance weights which clip around the wheel spokes are normally available in 20 or 30 gram sizes. If they are not available, wire solder, wrapped around the spokes close to the spoke nipples, forms a good substitute.

5  For machines fitted with cast alloy wheels, two weights of wheel balance weight may be obtained. Both types are designed so that they may be clipped to the wheel rim.

6  Although the rear wheel is more tolerant to out-of-balance forces than is the front wheel, ideally this too should be balanced if a new tyre is fitted. Because of the drag of the final drive components the chain must be removed from the rear sprocket. Balancing can then be carried out as for the front wheel.

22.1 Do not omit to refit the tyre valve dust cap

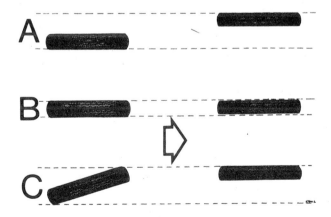

Fig. 5.8 Method of checking wheel alignment

*A & C – Incorrect*                    *B – Correct*

## Tyre changing sequence - tubed tyres

**A** Deflate tyre. After pushing tyre beads away from rim flanges push tyre bead into well of rim at point opposite valve. Insert tyre lever adjacent to valve and work bead over edge of rim.

**B** Use two levers to work bead over edge of rim. Note use of rim protectors

**C** Remove inner tube from tyre

**D** When first bead is clear, remove tyre as shown

**E** When fitting, partially inflate inner tube and insert in tyre

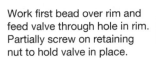

**F** Work first bead over rim and feed valve through hole in rim. Partially screw on retaining nut to hold valve in place.

**G** Check that inner tube is positioned correctly and work second bead over rim using tyre levers. Start at a point opposite valve.

**H** Work final area of bead over rim whilst pushing valve inwards to ensure that inner tube is not trapped

## 24 Fault diagnosis: wheels, brakes and tyres

| Symptom | Cause | Remedy |
| --- | --- | --- |
| Handlebars oscillate at low speed | Buckle or flat in wheel rim, most probably front wheel | Check rim for damage by spinning wheel. Renew wheel if not true. |
| | Tyre pressure incorrect | Check, and if necessary adjust. |
| | Tyre not straight on rim | Check tyre fitting. If necessary, deflate tyre and reposition. |
| | Worn wheel or steering head bearings | Check and renew or adjust. |
| Machine tends to weave | Tyre pressure incorrect | Check, and if necessary adjust. If sudden, check for puncture. |
| | Suspension worn or damaged | Check action of front forks and rear suspension units. Check swinging arm for wear. |
| Machine lacks power and accelerates poorly | Front brake binding | Drum brake models: Hot drum indicates binding. Check adjustment. Check drum for distortion. |
| | | Disc brake models: Hot disc or caliper indicates binding. Overhaul caliper and master cylinder, fit new pads if required, check disc for scoring or warpage. |
| | Rear brake binding | As for front drum. |
| Brakes grab of judder when applied gently | Disc brake: Pads badly worn or scored | Renew pads and check disc and caliper. |
| | Wrong type of pad fitted | Renew. |
| | Warped disc | |
| | Drum brake: Ends of brake shoes not chamfered | Chamfer with a file. |
| | Elliptical brake drums | Light skim in lathe by specialist. |
| Brake squeals | Disc brake: Glazed pads. Pads worn to backing metal | Sand pad surface to remove glaze then use brake gently for about 100 miles to permit bleeding in. If worn to backing check that disc is not damaged and renew as necessary. |
| | Caliper and pads polluted with brake dust or foreign matter | Dismantle and clean. Overhaul calipr where necessary. |
| | Drum brake: Glazed brake shoes | |
| | Shoes worn to backing metal | Remove glaze as above. |
| | Shoes and drum polluted as above | Dismantle and clean. |
| Excessive front brake lever travel | Disc brake: Air in system | Find cause of air's presence. If due to leak, rectify, then bleed brake. |
| | Very badly worn pads | Renew, and overhaul system where required. |
| | Badly polluted caliper | Dismantle and clean. |
| | Drum brake: Incorrect adjustment of cable | Readjust. |
| | Very badly worn brake shoes | Renew. |
| Front brake lever feels springy | Disc brake: Air in system | See above. |
| | Pads glazed | See above. |
| | Caliper jamming | Dismantle and overhaul. |
| | Drum brake: Pads glazed | See above. |

| Symptom | Cause | Remedy |
|---|---|---|
| Front and rear brake pull-off spongy (drum brakes only) | Brake cam binding in housing | Free and grease. |
| | Weak brake shoe springs | Renew if springs have not become displaced. |
| Harsh transmission | Worn or badly adjusted final drive chain | Adjust or renew. |
| | Hooked or badly worn sprockets | Renew as a pair. |
| | Loose rear sprocket | Check bolts. |
| | Worn damper rubbers | Renew rubber inserts. |
| Tyres wear more rapidly in middle of tread | Over-inflation | Check pressures and run at recommended settings. |
| Tyres wear rapidly at outer edge of tread | Under-inflation | As above. |

# Chapter 6 Electrical system

## Contents

## Specifications

### Battery

| | |
|---|---|
| Make ............................................................... | Yuasa |
| Type ............................................................... | 6N4-2A |
| Voltage ........................................................... | 6 volt |
| Capacity .......................................................... | 4 Ah |
| Electrolyte specific gravity ................................. | 1.26 at 20°C (68°F) |
| Earth .............................................................. | Negative |

### Fuse

| | |
|---|---|
| Rating ............................................................. | 10 amp |

### Resistor

| | |
|---|---|
| Rating ............................................................. | 4 ohm |

### Flywheel generator

| | GP100 | GP125 |
|---|---|---|
| Ignition source coil resistance ............................ | 0.1 ohm | 0.05 ohm |
| Charging coil resistance ................................... | 0.2 ohm | 0.2 ohm |
| Lighting coil resistance .................................... | 0.3 ohm | 0.2 ohm |
| Charging rate: | | |
|   Lights on: | | |
|     At 4000 rp, ............................................ | Above 0.6 amp | Above 0.7 amp |
|     AT 8000 rpm ......................................... | Below 2.8 amp | Below 1.5 amp |
|   Lights off: | | |
|     At 4000 rpm ......................................... | Above 0.7 amp | Above 0.8 amp |
|     At 8000 rpm ......................................... | Below 2.8 amp | Below 2.5 amp |
|   Lighting coil output: | | |
|     At 2500 rpm .......................................... | Above 5.7 volt | Above 6.0 volt |
|     At 8000 rpm .......................................... | Below 8.7 volt | Below 8.5 volt |

## Bulbs

| | |
|---|---|
| Headlamp | 6V, 25/25W |
| Tail/stop lamp | 6V, 5/21W |
| Direction indicators: | |
|     C suffix models | 6V, 10W |
|     N, X, D and L suffix models | 6V, 18W or 21W* |
| Speedometer light | 6V, 1.7W |
| Tachometer light | 6V, 1.7W |
| Main beam indicator | 6V, 1.7W |
| Neutral indicator | 6V, 3W |
| Flashing indicator warning light | 6V, 3W |
| Pilot lamp | 6V, 3W |

*check bulb holder marking and flasher unit rating to ensure correct bulb wattages are used*

## 1 General description

The Suzuki GP100 and GP125 models covered in this Manual are all fitted with a flywheel generator which contains two separate power coils; one to provide ignition source power and the other to provide power for the lighting system. These two coils are in no way connected and for the purposes of testing and fault isolating may be considered as separate component systems.

The lighting coil of the flywheel generator produces alternating current (ac) which must be converted to direct current (dc) to make it compatible with the battery and components of the electrical system.

This is achieved by means of a silicon diode rectifier, which effectively blocks half of the output wave by acting as a oneway electronic switch. For obvious reasons, this system is known as half-wave rectification. The resulting dc current is used to power the electrical system and to charge the 6 volt 4 amp hour battery. The electrical system is protected by a 10 amp fuse which is incorporated in the positive lead to the battery.

## 2 Testing the electrical system: general information

1 The electrical system incorporated in the Suzuki GP100 and GP125 models lends itself easily to fairly comprehensive testing of its component parts. A certain amount of preliminary dismantling will be necessary to gain access to the components to be tested. Normally, removal of the side panels and headlamp reflector unit will be required, with the possible addition of the seat and fuel tank.

2 Simple continuity checks may be made using a dry battery and bulb arrangement, but for most of the tests in this Chapter a pocket multimeter can be considered essential. Many owners will already possess one of these devices, but if necessary they can be obtained from electrical specialists, mail order companies or can be purchased from a Suzuki service agent as a 'pocket tester', part number 09900 - 25002.

3 Care must be taken when performing any electrical test, because some of the electronic assemblies can be destroyed if they are connected incorrectly or inadvertently shorted to earth. Instructions regarding meter probe connections are given for each test, and these should be read carefully to preclude any accidental damage during the test. Note that separate amp, volt and ohm meters may be used in place of the multimeter if necessary, noting that the appropriate test ranges will be required.

4 Where test equipment is not available, or the owner feels unsure of the procedure described, it is recommended that professional assistance is sought. Do not forget that a simple error can destroy a component such as the rectifier, resulting in expensive replacements being necessary.

## 3 Wiring: layout and examination

1 The wiring harness is colour-coded and will correspond with the accompanying diagrams. Where socket connectors are used they are designed so that reconnection can be made only in the one correct position.

2 Visual inspection will show whether any breaks or frayed outer coverings are giving rise to short circuits. Another source of trouble may be the snap connectors and sockets, where the connector has not been pushed home fully in the outer housing.

3 Intermittent short circuits can often be traced to a chafed wire that passes through or is close to a metal component, such as a frame member. Avoid tight bends in the wire or situations where the wire can become trapped between casings.

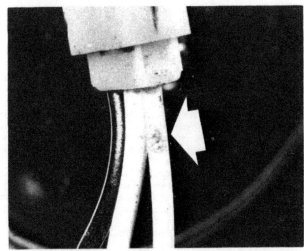

3.3 Avoid the possibility of wires chafing on metal components

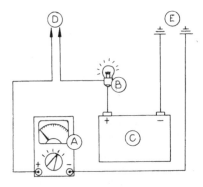

**Fig. 6.1 Simple testing arrangement for checking the wiring**

| | | |
|---|---|---|
| A   Multimeter | C   Battery | E   Negative |
| B   Bulb | D   Positive | |

## 4  Flywheel generator: checking the output

### Charging coil performance

1    The purpose of this test is to check the amount of electrical current being fed to the battery from the charging coil of the flywheel generator. This test should be carried out at the positive (+) connection of the battery so that the amount of direct current (dc) flowing from the voltage rectifier unit is measured.

2    Commence the test by unplugging the positive battery connection. Ensure that the two halves of the connection are clean and connect a multimeter between them. Set the meter on its dc ampere range (0-20 amps), check that it is properly supported so that its connections will not become detached from the battery leads and then start the engine. The readings obtained in the meter scale should correspond with those given below.

| Model type | Engine speed | Lighting state | Charging current |
|---|---|---|---|
| GP100 | 4000 rpm | On | 0.6 amp min |
|  | 8000 rpm | On | 2.8 amp max |
|  | 4000 rpm | Off | 0.7 amp min |
|  | 8000 rpm | Off | 2.8 amp max |
| GP125 | 4000 rpm | On | 0.7 amp min |
|  | 8000 rpm | On | 1.5 amp max |
|  | 4000 rpm | Off | 0.8 amp min |
|  | 8000 rpm | Off | 2.5 amp max |

### Lighting coil performance

3    The purpose of this text is to check the amount of voltage being fed to the headlamp bulb from the lighting coil of the flywheel generator. Prepare for the test by removing the headlamp reflector unit. Trace the yellow lead which runs from the headlamp dip switch to the headlamp bulb holder and disconnect it by pulling apart the two halves of the block connector through which it passes.

4    Set a multimeter to its 0 - 10 ac volt range and connect its positive probe to the yellow wire terminal. Connect the negative probe of the meter to a good earth point, set the switch button to the 'Hi' position and start the engine. The readings obtained on the meter scale should correspond with those given below.

| Model type | Engine speed | Voltage reading |
|---|---|---|
| GP100 | 2500 rpm | 5.7 volt min |
|  | 8000 rpm | 8.7 volt max |
| GP125 | 2500 rpm | 6.0 volt min |
|  | 8000 rpm | 8.5 volt max |

### Elimination checks

5    If the results obtained in either one of the above tests are unsatisfactory, then it is possible that the coil in question has failed and will need renewing. Before assuming that this is the case, carry out the following checks.

6    Refer to the wiring diagram at the end of this Chapter for the machine in question and carry out a check for continuity on the wires running between the coil and the point in the circuit at which the test was carried out. Separate each block or bullet connector in the circuit and check that it is free of all traces of dirt, moisture and corrosion.

7    Visually check each wire for signs of its having chafed against an engine or frame component and check any component contained within the circuit for serviceability by following the procedure given in the relevant Section of this Chapter. Finally, the condition of the coil in question can be checked by determining whether continuity exists.

### Continuity checks

8    To check any one coil for continuity, trace the wires from the flywheel generator stator to their nearest connection points

and separate each connection. Set a multimeter to its resistance function and connect it as shown in the table below:

| Model type | Coil | Connection | Resistance |
|---|---|---|---|
| GP100 | Ignition source | Black/Yellow to earth | 0.1 ohm |
|  | Charging | White/red to earth | 0.2 ohm |
|  | Lighting | Yellow/white to earth | 0.3 ohm |
| GP125 | Ignition source | Black/yellow to earth | 0.05 ohm |
|  | Charging | White/red to earth | 0.2 ohm |
|  | Lighting | Yellow/white to earth | 0.2 ohm |

If the readings shown on the meter scale do not correspond with those given above, then the coil in question is unserviceable and must be renewed.

## 5  Rectifier unit: location and testing

1    The rectifier unit is mounted in front of the battery, access being made by removing the left-hand side panel. The function of this unit is to convert the alternating current (ac) from the flywheel generator to direct current (dc) which can then be used to charge the battery. This it does by offering a very low resistance to current flow in one direction and an extremely high resistance in the reverse direction. As the ratios of these resistances are very high, an effective current will flow in one direction only.

2    The rectifier unit is mounted in such a way that it is not exposed to direct contamination by road salts, water or oil, yet has free circulation of air to permit cooling. It should be kept clean and dry. The unit should not give trouble during normal service. It can however be damaged by inadvertently reversing the battery connections.

3    To test the unit for continuity, set the multimeter to its resistance function (X1 ohm range) and connect its positive (+) probe to the negative (male) connection of the unit. Connect the negative (-) probe to the positive (female) connection of the unit and note the reading on the meter scale.

4    Reverse the position of the meter probes and once again note the meter reading. If an indication of continuity is given by the first meter reading and an indication of non-continuity shown by the second, then the unit is serviceable.

5    The test can, of course be carried out by connecting the rectifier unit to a simple battery and bulb test circuit. If the bulb lights with the circuit leads connected in one position and fails to light when the leads are reversed, then this indicates that the unit is doing its intended task, namely, allowing the current to flow in one direction only.

## 6  Ballast resistor: function and location

1    A ballast resistor is incorporated in the charging system to help prevent overcharging of the battery when the lights are not in use.

2    The resistor is fitted within a sealed box below and to the rear of the fuel tank retaining bolt. If removal is required, disconnect the two wires at their snap connectors and remove the fasteners.

3    If overcharging is experienced, indicated by a rapid level drop of the battery electrolyte, or if poor charging occurs when the lights are off, there is evidence that the resistor has failed. A multimeter, set to its x 1 ohm resistance range can be used to test the resistor. Disconnect the resistor and measure the resistance across its two wires; if the reading obtained differs significantly from the specified resistance of 4 ohms the unit is faulty and must be renewed.

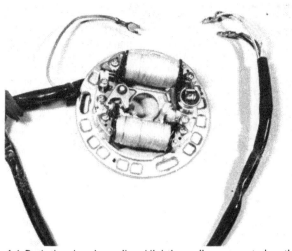

4.1 Both the charging coil and lighting coil are mounted on the flywheel generator stator

5.1 The rectifier is located just to the rear of the battery

6.1 The ballast resistor is mounted beneath the fuel tank mounting bolt

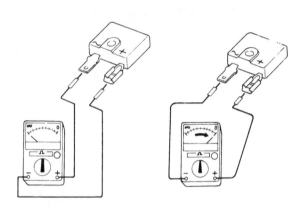

Fig. 6.2 Rectifier resistance test

### 7 Battery: examination and maintenance

1   A Yuasa 6N4-2A battery is fitted as standard to all of the models covered in this Manual and is of the lead-acid type. The translucent plastic case of the battery permits the upper and lower levels of the electrolyte to be observed when the battery is exposed by removal of the right-hand side panel from the machine. Maintenance is normally limited to keeping the electrolyte level between the prescribed upper and lower limits and by making sure the vent pipe is not blocked and remains correctly routed.

2   Unless acid is spilt, as may occur if the machine falls over, the electrolyte should always be topped up with distilled water, to restore the correct level. If acid is spilt on any of the machine, it should be neutralised with an alkali such as washing soda and washed away with plenty of water, otherwise serious corrosion will occur. Top up with sulphuric acid of the correct specific gravity (1.260) only when spillage has occurred. Check that the vent pipe is well clear of the frame tubes or any of the other cycle parts, for obvious reasons.

3   If battery problems are experienced, the following checks will determine whether renewal is required. A battery can normally be expected to last for about 3 years, but this life can be shortened dramatically by neglect. In normal use, the capacity for storage will gradually diminish, and a point will be reached where the battery will no longer hold adequate charge.

4   Remove the flat battery and examine the cell and plate condition near the bottom of the casing. An accumulation of white sludge around the bottom of the cells indicates sulphation, a condition which indicates the imminent demise of the battery. Little can be done to reverse this process, but it may help to have the electrolyte drained, the battery flushed and then refilled with new electrolyte. Most electrical wholesalers have facilities for this work.

5   Warping of the plates or separators is also indicative of an expiring battery, and will often be evident in only one or two of the cells. It can often be caused by old age, but a new battery which is overcharged will show the same failure. There is no cure for the problem, and the need to avoid overcharging cannot be overstressed.

6   Try charging the suspect battery as described in the following Section. If the battery fails to accept a full charge, and in particular if one or more cells show a low hydrometer reading, the battery is in need of renewal.

7   A hydrometer will be required to check the specific gravity of the electrolyte, and thus the state of charge. Any small hydrometer will do, but avoid the very large commercial types because there will be insufficient electrolyte to provide a

reading. When fully charged, each cell should read 1.260, with little discrepancy between cells.

8  Note that it is seldom practicable to repair a cracked battery case because the acid present in the joint will prevent the formation of an effective seal. It is always best to renew a cracked battery, especially in view of the corrosion which will be caused if the acid continues to leak.

9  If the machine has remained unused for a period of time, it is advisable to remove the battery and give it a refresher charge every six weeks or so from a battery charger. If the battery is permitted to discharge completely, the plates will sulphate and render the battery useless.

10 Occasionally, check the condition of the battery connections to ensure that they are free of corrosion and forming a good contact. If corrosion has occurred, it should be cleaned away by scraping with a knife and then using emery cloth to remove the final traces. Remake the electrical connections whilst the joint is still clean, having first smeared them lightly with petroleum jelly (not grease) to prevent recurrence of the corrosion. Badly corroded connections can have a high electrical resistance and may give the impression of a complete battery failure.

7.1 Observe the battery electrolyte level

7.2 Avoid trapping the battery vent pipe between frame components

## 8  Battery: charging procedure

1  The normal charging rate for the battery fitted to the machines covered in this Manual is 0.4 amp. It is permissible to charge at a more rapid rate in an emergency but this effectively shortens the life of the battery and should therefore be avoided. Because of this, go for the smallest charging rate available. Avoid quick charge services offered by garages. This will indeed charge the battery rapidly but it will also overheat it and may halve its life expectancy.

2  Never omit to remove the battery cell caps or neglect to check that the side vent is clear before recharging a battery, otherwise the gas created within the battery when charging takes place might burst the case with disastrous consequences. Do not attempt to charge the battery with it in-situ and with the leads still connected. This can lead to failure of the rectifier unit.

3  Make sure that the battery charger connections are correct; red to positive and black/white to negative. When the battery is reconnected to the machine, the black/white lead must be connected to the negative terminal and the red lead to positive. This is most important, as the machine has a negative earth system. If the terminals are inadvertently reversed, the electrical system will be damaged permanently. The rectifier unit can be destroyed by a reversal of the current flow.

4  A word of caution concerning batteries. Sulphuric acid is extremely corrosive and must be handled with great respect. Do not forget that the outside of the battery is likely to retain traces of acid from previous spills, and the hands should always be washed promptly after checking the battery. Remember too that battery acid will quickly destroy clothing.

5  Note the following rules concerning battery maintenance:
  Do not allow smoking or naked flames near batteries.
  Do avoid acid contact with skin, eyes and clothing.
  Do keep battery electrolyte level maintained.
  Do avoid over-high charge rates.
  Do avoid leaving the battery discharged.
  Do avoid freezing.
  Do use only distilled or demineralised water for topping up.

## 9  Fuse: location, function and renewal

1  The fuse is contained within a plastic holder which is clipped to a point just forward of the battery. A spare fuse of the same, 10 amp, rating is supplied with the machine when new and is contained within a clear plastic holder which is attached to the wire leading from the fuse in circuit.

2  The fuse is fitted to protect the electrical system in the event of a short circuit or sudden surge. It is, in effect, an intentional 'weak link' which will blow in preference to the circuit burning out.

3  Before replacing a fuse that has blown, check that no obvious short circuit has occurred, otherwise the replacement fuse will blow immediately it is inserted. It is always wise to check the electrical circuit thoroughly, to trace the fault and eliminate it.

4  When a fuse blows while the machine is running and no spare is available, a 'get you home' remedy is to remove the blown fuse and wrap it in silver paper before replacing it in the fuseholder. The silver paper will restore the electrical continuity by bridging the broken fuse wire. This expedient should **never** be used if there is evidence of a short circuit or other major electrical fault, otherwise more serious damage will be caused. Replace the 'doctored' fuse at the earliest possible opportunity to restore full circuit protection. It follows that spare fuses that are used should be replaced as soon as possible to prevent the above situation from arising.

## 10 Headlamp: bulb renewal and beam alignment

1   In order to gain access to the headlamp bulbs, it is necessary to remove the headlamp rim, complete with the reflector and headlamp glass. The rim is retained by two (or three) crosshead screws, each with a spring washer and a plain washer fitted beneath its head, which pass through the headlamp shell and into a threaded projection from the rim edge. With those screws removed, the rim may be drawn out of the headlamp shell.

2   The main headlamp bulb is a push fit in the central bulb holder of the reflector. This holder can be fitted in one position only to ensure that the bulb is always correctly focussed; it has a fitting of the bayonet type. A twin filament bulb of a 6 volt, 25/25 watt rating is used. This bulb is released from the holder by pressing inwards, turning anti-clockwise and then pulling it outwards.

3   The holder for the pilot lamp bulb is a direct push fit into the reflector. The bulb is of a 6 volt, 3 watt rating and has a bayonet fitting in the holder.

4   Beam height is effected by tilting the headlamp shell, after the mounting bolts have been loosened slightly. No provision is made for adjusting the horizontal alignment of the beam.

5   In the UK, regulations stipulate that the headlamp must be arranged so that the light will not dazzle a person standing at a distance greater than 25 feet from the lamp, whose eye level is not less than 3 feet 6 inches above that plane. It is easy to approximate this setting by placing the machine 25 feet away from a wall, on a level road, and setting the dip beam height so that it is concentrated at the same height as the distance of the centre of the headlamp from the ground. The rider must be seated normally during this operation and also the pillion passenger, if one is carried regularly.

6   Most other areas have similar regulations controlling headlamp beam alignment, and these should be checked before any adjustment is made.

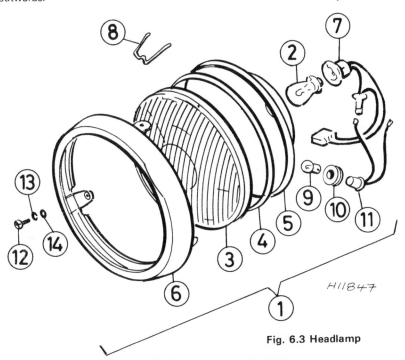

1   Headlamp assembly
2   Bulb
3   Glass
4   Sealing ring
5   Reflector
6   Rim
7   Bulb holder
8   Spring clip – 3 off
9   Pilot lamp bulb
10  Grommet
11  Pilot lamp bulb holder
12  Screw
13  Spring washer
14  Washer

Fig. 6.3 Headlamp

9.1 The fuse is contained within a plastic holder

10.1 Detach the headlamp rim ...

10.2a ... pull out the headlamp bulb holder ...

10.2b ... and remove the headlamp bulb

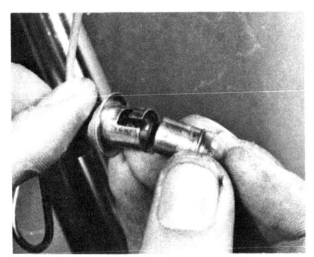

10.3 The pilot lamp bulb is a bayonet fit in its holder

## 11 Stop and tail lamp: bulb renewal

1    The combined tail and stop lamp unit fitted to these models is fitted with a double filament bulb having offset pins to prevent its unintentional reversal in the bulb holder. The lamp unit serves a two-fold purpose – to illuminate the rear of the machine and the rear number plate, and to give visual warning when the rear brake is applied.

2    To gain access to the bulb, remove the two screws with sealing washers which hold the red plastic lens in position. The bulb is released by pressing inwards and with an anti-clockwise turning action, the bulb will now come out. When lifting the lens from the backplate, take care not to tear the sealing gasket located between the two. This gasket is essential to keep moisture and dirt from entering the electrical contacts within the unit and to stop corrosion or dulling of the backplate. The sealing washers on the lens securing screws serve a similar function and should be renewed if seen to be damaged.

3    Refitting of the bulb and lens is a reversal of the removal procedure. Check the inside of the bulb holder for any signs of corrosion or moisture, ensuring that the contacts are free to move when depressed. When refitting the lens securing screws, take care not to overtighten them as it is possible that the lens may crack.

11.3a The stop/tail lamp bulb is a bayonet fit in its holder

11.3b Ensure the lens seal is serviceable

11.3c Do not overtighten the lens securing screws

## 12 Flashing indicator lamps: bulb renewal

1   Flashing indicator lamps are fitted to the front and rear of the machine. They are mounted on short stalks through which the wires pass. Access to each bulb is gained by removing the two screws holding the plastic lens cover.
2   The bulbs fitted are of the bayonet type and may be released by pushing in, turning anti-clockwise and pulling out of the holder.
3   Refitting of the bulb and lens is a reversal of the removal procedure. Check the inside of the bulb holder for any signs of corrosion or moisture, ensuring that the contact is free to move when depressed. Do not omit to fit a lens seal which is in good condition and take great care not to overtighten the lens securing screws as it is possible to crack the lens by doing so.

## 13 Instrument console assembly: bulb renewal

1   Access may be gained to the various warning and instrument illumination bulbs by removing the bottom plate from the instrument console and unplugging the various bulb holders from the base of each instrument head.
2   The bottom plate of the console can be detached by removing the two dome nuts, each with its plate washer, from the base of each instrument. Once detached, the plate can be lowered down over the drive cables until the bulb holders are fully exposed.
3   Each bulb holder is a straight push-fit in the base of its respective instrument head. With the holder pulled from position, its bulb may be removed by pushing it inwards, turning it anti-clockwise and then pulling it out of the holder.
4   Before fitting the replacement bulb, check that the contact within the holder is free to move when depressed and remove any signs of moisture or corrosion from within the holder. Fitting the bulb and holder and reassembling the console is a direct reversal of the above listed procedure.
5   Later GP100 UD and all UL models are fitted with capless bulbs in the instrument panel. These are a press fit in their respective holders; take care not to damage their fine wire tails on refitting.

## 14 Ignition switch: location and testing

1   The ignition switch is situated on the instrument console, directly above the steering head upper yoke. It may be tested for continuity at the block connector within the headlamp shell.
2   Remove the headlamp reflector unit and identify the block connector by the red, orange, black/white and black/yellow wires running into it from the switch. Disconnect the switch by pulling apart the two halves of the connector.

3   With the switch turned to the 'On' position, check for continuity between the red and the orange wire terminals. With the switch turned to the 'Off position', check for continuity between the black/yellow and the black/white wire terminals.
4   If continuity is found to exist in both of the aforementioned tests, then the switch is serviceable. If either one of the tests shows non-continuity, then the switch is unserviceable and should be renewed. Before removing the switch for renewal, carry out a physical check of the wiring from the switch to the block connector to ensure that there is no indication of the wires having been cut or frayed by any of the associated cycle parts.
5   The switch may be detached from the instrument console by unscrewing the retaining ring at the top of the switch and pulling the switch body down out of its location.

## 15 Neutral indicator switch: location and testing

1   The neutral indicator switch takes the form of a white plastic cover which is sited over the left-hand end of the gearchange selector drum. Access to the switch may be gained by removing the rearmost section of the left-hand crankcase cover.
2   To test the switch, set a multimeter to its resistance function and carry out a check for continuity between the switch terminal and earth with the gearchange lever set in the neutral position. If continuity is found, then the switch is serviceable. If this is not the case, then the switch should be removed by following the instructions given in Section 9 of Chapter 1. Full instructions for fitting a switch are given in Section 41 of Chapter 1.

## 16 Brake stop lamp switches: adjustment and testing

### Rear brake switch

1   The rear brake stop lamp switch is located on a frame-mounted bracket which is situated directly above the pivot for the swinging arm on the right-hand side of the machine. It is operated by movement of the rear brake pedal, the two components being adjoined by an extension spring. The body of the switch is threaded to permit adjustment and is secured to the bracket by two nuts.
2   If the stop lamp is seen to be late in operating, raise the switch body by first loosening the two nuts and then rotating them in a clockwise direction whilst holding the switch steady. Once the switch is set in the required position, retain it by tightening the two nuts against each other. If the stop lamp is seen to be permanently on, then the switch body should be lowered in relation to its mounting bracket. As a guide to operation, the stop lamp should illuminate immediately after the brake pedal is depressed.
3   Testing the switch is a simple matter of checking for continuity between the switch terminals with the switch fully extended, that is in the 'On' position. Before testing, check the switch adjustment by referring to the above text. Disconnect the two electrical leads from the switch terminals. Set a multimeter to its resistance function and with the probes of the meter connected one to each terminal, operate the switch. If no continuity is found, then the switch must be renewed as it is not possible to effect a satisfactory repair.

### Front brake switch

4   Some models are also fitted with a stop lamp switch on the front brake. The switch is attached to the lever pivot by two screws and can be adjusted, as described above, after the mounting screws have been slackened; be careful not to overtighten the screws on completion or the switch body may be cracked. Testing of the switch can be carried out as described in paragraph 3.

12.1 Detach each indicator lens ...

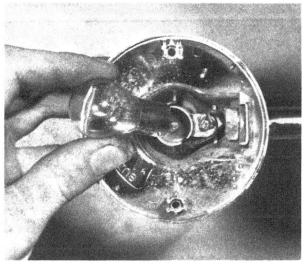

12.2 ... to gain access to the indicator bulb

13.2 Detach the instrument console bottom plate ...

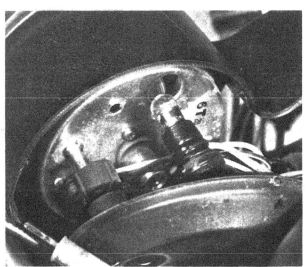

13.3 ... and pull the instrument bulb holders from the base of each instrument head

15.1 The neutral indicator switch is located beneath the gearbox sprocket

16.1 The rear brake stop lamp switch is mounted above the swinging arm pivot

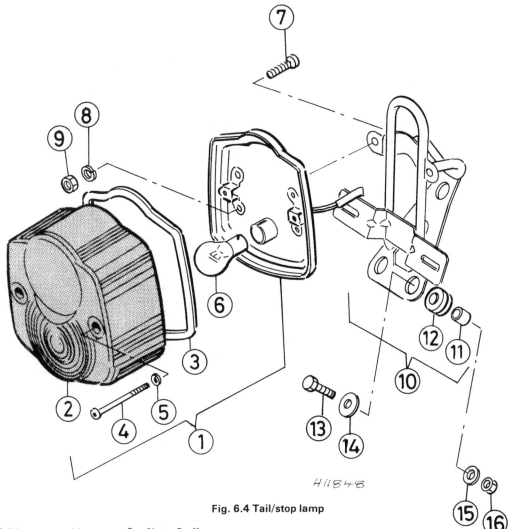

Fig. 6.4 Tail/stop lamp

| | | | |
|---|---|---|---|
| 1 | Tail lamp assembly | 9 | Nut – 2 off |
| 2 | Lens | 10 | Mounting bracket |
| 3 | Rubber seal | | assembly |
| 4 | Screw – 2 off | 11 | Spacer – 3 off |
| 5 | Sealing washer– | 12 | Grommet – 3 off |
| | 2 off | 13 | Bolt – 3 off |
| 6 | Stop/tail lamp bulb | 14 | Washer – 3 off |
| 7 | Screw – 2 off | 15 | Washer – 3 off |
| 8 | Spring washer– | 16 | Nut – 3 off |
| | 2 off | | |

## 17 Horn: location, adjustment and replacement

1   The horn is located beneath the headlamp unit and is secured to the steering head lower yoke by means of a flexible steel bracket. This bracket is secured to the yoke by two bolts and helps isolate the horn from high frequency vibration.

2   The horn should not normally require adjustment. If, however, this becomes necessary, then adjustment is provided by means of a screw at the rear of the horn case. This screw should be turned fractionally clockwise to increase the horn volume.

3   If necessary, the horn may be removed from the machine by disconnecting the two electrical wires from the horn terminals, after having first noted their fitted positions, and then detaching either the horn unit from the bracket or the bracket from the lower yoke. Fitting a horn unit is a direct reversal of the removal procedure.

17.1 The horn is mounted on the steering head lower yoke

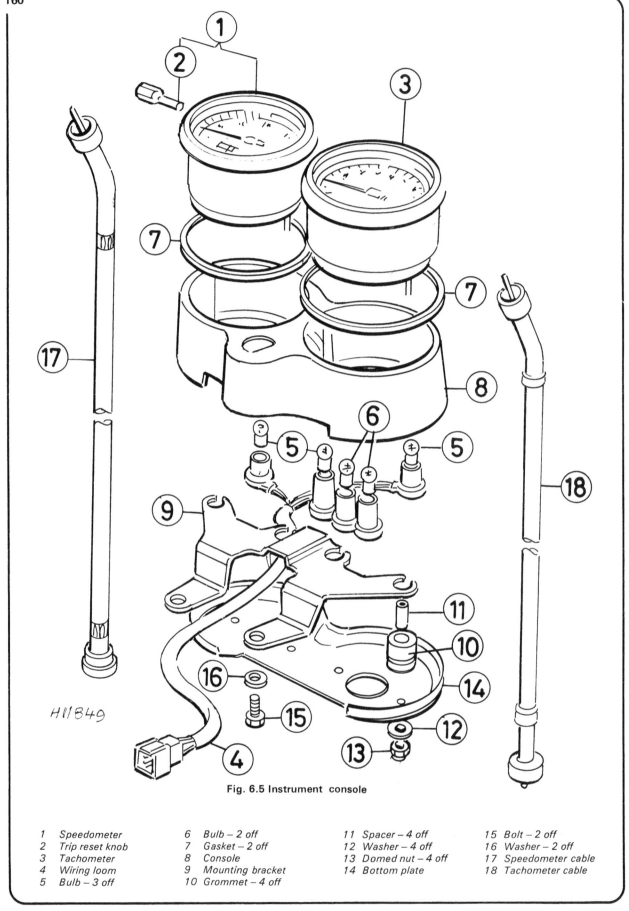

H11849

Fig. 6.5 Instrument console

| | | | | | |
|---|---|---|---|---|---|
| 1 | Speedometer | 6 | Bulb – 2 off | 11 | Spacer – 4 off |
| 2 | Trip reset knob | 7 | Gasket – 2 off | 12 | Washer – 4 off |
| 3 | Tachometer | 8 | Console | 13 | Domed nut – 4 off |
| 4 | Wiring loom | 9 | Mounting bracket | 14 | Bottom plate |
| 5 | Bulb – 3 off | 10 | Grommet – 4 off | | |

| | |
|---|---|
| 15 | Bolt – 2 off |
| 16 | Washer – 2 off |
| 17 | Speedometer cable |
| 18 | Tachometer cable |

## 18 Flasher unit: location and renewal

1    The flasher relay unit is located behind the left-hand side panel, just forward of the oil tank, and is supported on an anti-vibration mounting.

2    If the flasher unit is functioning correctly, a series of audible clicks will be heard when the indicator lamps are in operation. If the unit malfunctions and all the bulbs are in working order, the usual symptom is one initial flash before the unit goes dead; it will be necessary to replace the unit complete if the fault cannot be attributed to any other cause.

3    If renewal is necessary, then removal of the existing unit is simple. Unplug the block connector from the base of the unit and move the unit upwards to unhook it from its mounting. When fitting the new unit, take great care not to subject it to any sudden shock; it is easily damaged if dropped.

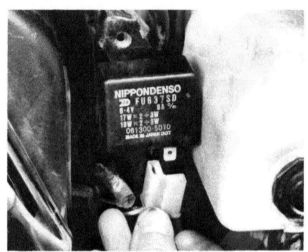

18.3a Unplug the block connector from the flasher unit ...

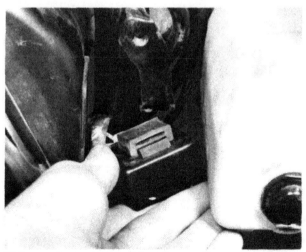

18.3b ... and unhook the unit from its mounting

## 19 Handlebar switches: general information

1    Generally speaking, the switches give little trouble, but if necessary they can be dismantled by separating the halves which form a split clamp around the handlebars.

2    Always disconnect the battery before removing any of the switches, to prevent the possibility of a short circuit. Most troubles are caused by dirty contacts, but in the event of the breakage of some internal part, it will be necessary to renew the complete switch.

3    Because the internal components of each switch are very small, and therefore difficult to dismantle and reassemble, it is suggested a special electrical contact cleaner be used to clean corroded contacts. This can be sprayed into each switch, without the need for dismantling.

4    It will be necessary to obtain a multimeter and set it to its resistance function in order to carry out the various continuity checks described in the following Sections.

## 20 Lighting switch: testing

1    The lighting switch is incorporated in the upper half of the switch assembly located adjacent to the left-hand handlebar grip rubber.

2    Remove the headlamp reflector unit and identify the green/white, white/red, yellow/white and grey wires which run from the switch to its block connector. Disconnect the switch by pulling apart the two halves of the connector.

3    To test the switch, move the switch button to the 'Off' position and check for continuity between the green/white and white/red wire terminals. Move the switch button to the 'On' position and check for continuity between the white/red and yellow/white wire terminals, the yellow/white and grey wire terminals and the white/red and grey wire terminals.

4    If continuity is found to exist in all of the tests, then the switch is serviceable. If either one of these tests shows non-continuity, then the switch must be renewed. Remember that any failure that occurs within the switch can plunge the lighting system into darkness, with disastrous consequences.

## 21 Headlamp dip switch: testing

1    The headlamp dip switch is incorporated in the upper half of the switch assembly located adjacent to the left-hand handlebar grip rubber.

2    Remove the headlamp reflector unit and identify the grey, white and yellow wires which run from the switch to its block connector. Disconnect the switch by pulling apart the two halves of the connector.

3    To test the switch, move the switch button to the 'Hi' position and check for continuity between the grey and the yellow wire terminals. Move the switch button to the 'Lo' position and check for continuity between the grey and the white wire terminals.

4    If continuity is found to exist in both of the aforementioned tests, then the switch is serviceable. If either one of these tests shows non-continuity then the switch must be renewed, as any failure that occurs when changing from one beam to the other can plunge the lighting system into darkness, with disastrous consequences.

## 22 Flashing indicator switch: testing

1    The direction indicator switch is incorporated in the switch assembly located adjacent to the left-hand handlebar grip rubber.

2    Remove the headlamp reflector unit and identify the black, light blue and light green wires which run from the switch to its block connector. Disconnect the switch by pulling apart the two halves of the connector.

3    To test the switch, move the switch button to the 'R' position and check for continuity between the light blue and the light green wire terminals. Move the switch button to the 'L'

position and check for continuity between the black and the light blue wire terminals.

4   If continuity is found to exist in both of the aforementioned tests, then the switch is serviceable. Otherwise, the switch is defective and will need to be examined and, if necessary, renewed.

## 23 Horn switch: testing

1   The horn switch takes the form of a push button which is incorporated in the lower half of the switch assembly located adjacent to the left-hand handlebar grip rubber.

2   To test the switch, remove the headlamp reflector unit and trace the green wire from the switch to its block connector. Separate the two halves of this connector and carry out a test for continuity between the wire terminal and an earth point of the handlebar with the button fully depressed.

3   If continuity is seen to exist, then the switch is serviceable. Otherwise, the switch is defective and will need to be examined and, if necessary, renewed.

## 24 Engine stop switch: testing

1   The engine stop switch is incorporated in the switch unit located adjacent to the throttle twistgrip. To test the switch, first remove the headlamp reflector unit and trace the black/yellow wire from the switch to its block connector. Separate the two halves of the connector and carry out a test for continuity between the wire terminal and an earth point of the handlebar with the switch set in the 'Run' position.

2   If a reading of non-continuity is given, then the switch is serviceable. Otherwise, the switch is defective and will need to be examined and, if necessary, renewed.

## 25 Fault diagnosis : electrical system

| Symptom | Cause | Remedy |
| --- | --- | --- |
| Complete electrical failure | Blown fuse | Check wiring and electrical components for short circuit before fitting new 10 amp fuse. Check battery connections, also whether connections show signs of corrosion. |
| Dim lights, horn inoperative | Discharged battery | Re-charge battery with battery charger. Check whether generator is giving correct output. Check rectifier. |
| Constantly blowing bulbs | Vibration, poor earth connection | Check security of bulb holders. Check earth return connections. |
|  | Damaged rectifier | Renew rectifier. |
| Persistent overcharging of battery | Damaged rectifier | Test and renew. |

# Chapter 7  The 1985 to 1986 GP125 N model

## Contents

### Introduction

The main UK market machines are covered in Chapters 1 to 6 of this manual. In addition to these models, however, an additional number of GP 125 N models were imported into the UK in 1985, and it is to these that this update Chapter relates.

To summarise, the machines covered by this manual as of August 1986 are as follows. The last letter of the model name is the manufacturer's model year code and this will be found on the back cover of the owners handbook supplied with the machine. It will be noted that the year code relates to the year in which the machine was first produced. In practice, many popular models continue to be sold for several years without change. In the following model summary, the discontinuation dates are for guidance only; most models were available from dealers for some time after this date.

| Model name | Introduced | Discontinued |
|---|---|---|
| GP100 UN | February 1980 | September 1982 |
| GP100 UX | September 1982 | June 1983 |
| GP100 UD | June 1983 | April 1991 |
| GP100 UL | April 1991 | Current |
| GP100 C | August 1978 | August 1980 |
| GP100 N | August 1980 | March 1981 |
| GP100 X | March 1981 | March 1984 |
| GP100 ED | March 1983 | 1986 |
| GP125 C | February 1978 | October 1979 |
| GP125 N | October 1979 | July 1982 |
| GP125 X | July 1982 | September 1983 |
| GP125 D | September 1983 | February 1989 |
| GP125 N | 1985 Import | 1986 |

The GP125 N appears twice in the above list; firstly in its original run between 1979 -- 82, and then in its 1985 -- 1986 guise. This update relates to the 1985 to 1986 imports of the GP125 N model where it differs from the original version. In other respects it is covered in Chapter 1 to 6. As an aid to identification, the later import versions of the GP125 N model are distinguished by a rectangular tail lamp assembly in place of the rounder unit fitted to all other UK models. The frame numbers of the imports lie between 140176 and 144954, but note that there is an overlap in these numbers and those of the original GP125 N model.

### 2  Ordering spare parts

1  As with all models, it is important to quote the engine and frame numbers in full when ordering replacement parts from a Service Agent. This is particularly relevant in the case of the 1985/1986 GP125 N model because of detail differences between it and the original model. If this precaution is not observed, there is a real risk that the new parts may not fit. As a guide, the following items are not interchangeable with those fitted to the original GP125 N.

Main bearings
Cylinder head
Throttle stop screw
Air filter assembly
Silencer
Fork stanchions
HT coil
Headlamp assembly
Reflectors
Generator stator
Generator rotor
Generator lighting coil
Generator primary coil
Ignition pulser coil
Rear lamp lens
Rear lamp base
Rear lamp bracket
Wiring harness
Decals

### 3  Ignition system: general

1  The single major difference between the 1985 on GP125 N models and the other models covered in this manual is the use of a capacitor discharge ignition (CDI) system, rather than a conventional flywheel magneto and contact breaker arrangement. The system is known by the initials PEI, or pointless electronic ignition. As the name indicates, the system dispenses with a contact breaker, or points, and is instead

triggered electronically. This means that no regular mainten-ance or adjustment is required, there being no mechanical parts to wear.

2   The system comprises a sealed CDI unit, powered from the low-speed and high-speed coils in the flywheel generator assembly. The two coils combine to supply a relatively constant voltage at all engine speeds. During each half revolution of the engine, one positive and one negative pulse is applied to the CDI unit. The positive pulse is used to supply a charge to the capacitor in the CDI unit, whilst the negative pulse supplies the trigger pulse to the timing circuit, and this is used to switch a device known as a silicon controlled rectifier (SCR) into a conductive state.

3   As the SCR becomes conductive, it allows the capacitor to discharge through the primary windings of the ignition coil. This in turn induces a high tension pulse in the secondary windings, and it is this which is discharged across the plug electrodes to provide the ignition spark. This results in two ignition sparks per engine revolution, only one of which is used to ignite the combustion mixture, the remaining spark being wasted. A further refinement of the CDI unit is a timing circuit. This gauges engine speed by counting the frequency of trigger pulses, and is thus able to correct the ignition timing electronically to suit various engine speeds. Two systems are employed, based either on Nippon Denso or Kokusan CDI units.

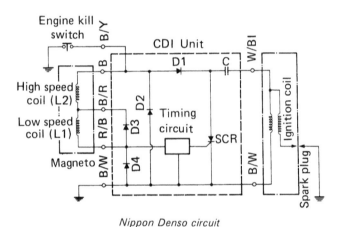

*Nippon Denso circuit*

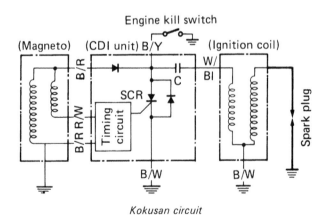

*Kokusan circuit*

**Fig. 7.1 PEI system circuit diagram**

| | | | |
|---|---|---|---|
| B | Black | W | White |
| Bl | Blue | Y | Yellow |
| R | Red | | |

### 4   Ignition system: fault diagnosis

1   Like most CDI systems, the Suzuki PEI arrangement will give long and trouble-free service, the only item requiring regular attention or renewal being the spark plug. As has been mentioned, there is no need for regular adjustment or timing checks as with contact breaker systems. In the event of a fault developing, the source of the problem should be tracked down in a logical sequence; checking components at random will almost invariably waste time and fail to locate the cause of the problem. Before attempting any diagnosis work, the following precautions should be noted: *Never turn the engine over without the HT lead connected to the spark plug and the plug securely earthed to the engine. If this precaution is not observed the CDI unit or the ignition coil can be irrevocably damaged. Electronic ignition systems generate very high voltages and shocks from the system can be very unpleasant or even dangerous. With a flywheel generator-powered system, the risk of such shocks is remote, but beware of pulling off the plug cap if the engine is running; even if you do not receive a shock the system may be damaged.* There is a limit to the degree of testing which can be carried out at home without specialist test equipment. Most tests can be carried out with an inexpensive pocket multimeter, available from motorcycle dealers or electrical stores. For a full and definitive test of the CDI unit, Service Agent facilities will be needed.

| Checks | Action |
|---|---|
| Spark plug | With the engine switched off, remove the spark plug and fit a new one, having set the electrode gap to the prescribed 0.6 -- 0.8 mm (0.024 -- 0.031 in) |
| System wiring and connections | Apart from plug faults, this is the single, most likely cause of problems. Look for loose, corroded or damaged connectors, water contamination (use WD40 or similar to displace water), frayed or broken leads. Repair or renew as required. Refer to the accompanying circuit diagram and the wiring diagram at the end of this Chapter for details. |
| Ignition and kill switches | Check operation of both switches, especially if the fault has developed while riding in wet weather. WD40 will displace water from switches. Packing switches with silicone grease will prevent similar problems recurring. |
| Ignition coil | Using a multimeter, check the ignition coil primary and secondary winding resistances. Renew coil if faulty. |
| Flywheel generator source coils | Using a multimeter, check stator coils for the correct resistances. Renew if faulty. |
| CDI unit | The CDI unit resistances can be given a rough check using a multimeter. A more comprehensive test requires specialist equipment unlikely to be available to the average owner. Alternatively, check by substituting a new unit. |

### 5   Ignition system: checking the ignition coil

1   The ignition coil is checked in much the same way as has been described in Section 6 of Chapter 3. The primary windings can be checked by measuring the resistance between the black/white lead and the white/blue lead (Nippon Denso coil) or between the black/yellow lead and the metal core of the coil (Kokusan coil).

2   Irrespective of the type of coil fitted, the secondary winding resistance measurement is made between the plug cap and the metal core of the coil. Check each resistance and compare the readings obtained with those shown below. Note that the readings shown are approximate only, but if the figures obtained are radically different, the coil can be assumed to be faulty and should be renewed.

| Resistance | Nippon Denso | Kokusan |
|---|---|---|
| Primary winding ................ | 0.5 ohm | 0.05 ohm |
| Secondary winding ............ | 13 K ohms | 12 K ohms |

### 6   Ignition system: checking the flywheel generator coils

This test is similar to that described for the contact breaker models in paragraph 8, Section 4 of Chapter 6. With the seat removed for access, trace back and disconnect the flywheel generator output leads at their connectors. Check with a multimeter between the various pairs of leads to establish the resistance figure of each coil, following the table shown below. Note that it is not strictly essential to test the charging and lighting coils, but that these may as well be checked at the same time. Note that the resistances shown below are approximate only, but if the readings obtained differ significantly, the coil(s) should be renewed.

| Coil | Wiring colours | Resistance |
|---|---|---|
| Nippon Denso type: | | |
| High speed ....... | Black to black/red | 27 ohms |
| Low speed ........ | Black/red to red/black | 200 ohms |
| Charging .......... | Green/white to earth | 0.2 ohms |
| Lighting ............ | Yellow to earth | 0.3 ohms |
| Kokusan type: | | |
| Exciter .............. | Black/red to black/white | 210 ohms |
| Pulser ............... | Black/red to red/white | 24 ohms |
| Charging .......... | Green/white to yellow | 0.2 ohms |
| Lighting ............ | Yellow to earth | 0.3 ohms |

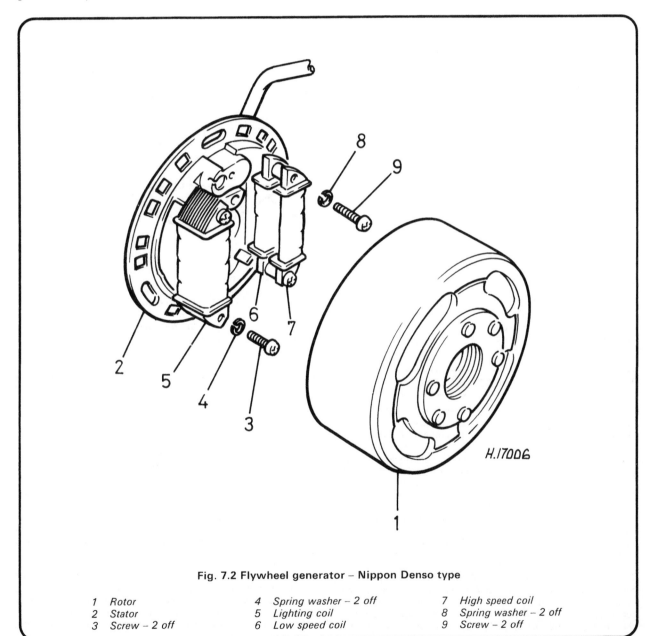

H.17006

**Fig. 7.2 Flywheel generator – Nippon Denso type**

| | | | | | |
|---|---|---|---|---|---|
| 1 | Rotor | 4 | Spring washer – 2 off | 7 | High speed coil |
| 2 | Stator | 5 | Lighting coil | 8 | Spring washer – 2 off |
| 3 | Screw – 2 off | 6 | Low speed coil | 9 | Screw – 2 off |

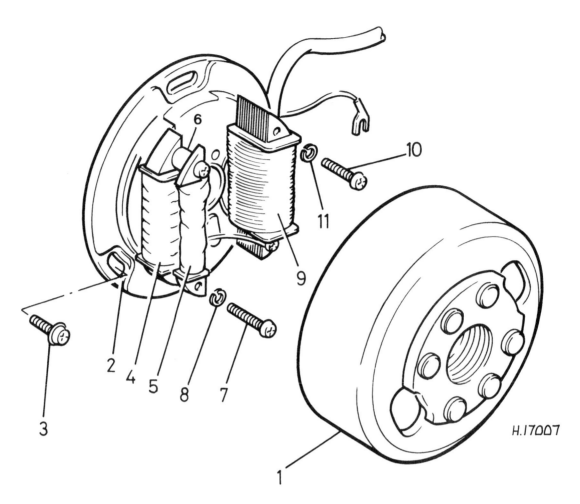

**Fig. 7.3 Flywheel generator – Kokusan type**

| | | |
|---|---|---|
| 1  Rotor | 5  Pulser coil | 9  Lighting coil |
| 2  Stator | 6  Spacer – 2 off | 10  Screw – 2 off |
| 3  Screw – 3 off | 7  Screw – 2 off | 11  Spring washer – 2 off |
| 4  Source coil | 8  Spring washer – 2 off | |

## 7  Ignition system: checking the CDI unit

1  If the elimination of other possible causes of an ignition fault has indicated that the CDI unit is suspect, it can be checked to some extent using a multimeter. As has been mentioned, a definitive test cannot be performed at home, but the resistance checks will usually suffice to confirm a fault. Start by removing the left-hand side panel to reveal the CDI unit. Remove the unit from the frame and place it on the workbench to await testing.

2  Before commencing the test, set the multimeter to the R x 1 K scale (kilo ohms) in the case of the Nippon Denso unit, and to the R x 100 (ohms x 100) scale for the Kokusan unit. Immediately before each stage of the test, connect a jumper lead across the pair of terminals to be checked to dissipate any residual charge in that part of the circuit.

3  Make a resistance check of each pair of leads, following the tables shown below. Note that ON indicates zero resistance, while OFF indicates infinite resistance. When comparing the readings obtained with those shown in the tables, note that they relate to the test as performed using the Suzuki Pocket Tester, Part Number 09900-25001. If using a different make of tester, some variation may be found. Finally, remember that this is not an exhaustive test; if a fault is revealed, have this confirmed by a full test at a Suzuki Service Agent before ordering a new unit.

## 8  Electrical system: modifications

1  As had been mentioned, there were a number of detail differences between the 1985 on import version of the GP125 N and the original model. Most commonly these relate to the fitting of slightly different electrical ancillaries, such as the headlamp and tail lamp assemblies. In the case of the headlamp, no parking light is fitted, and in consequence there is no setting for this provided on the ignition switch. The tail lamp shape was altered to a rectangular unit which was otherwise identical in operation. The front brake lamp switch was omitted.

2  The only significant consequence of these alterations was a revised wiring harness to suit the minor electrical system changes and the fitting of electronic ignition. The system is shown in full in the wiring diagram which will be found at the end of this Chapter.

| Put negative (−) pin of tester to: | Put positive (+) pin of tester to: | | | | | |
|---|---|---|---|---|---|---|
| | | B/Y | B | B/R | R/B | B/W | W/Bl |
| | B/Y | | ON | OFF | OFF | OFF | OFF |
| | B | ON | | OFF | OFF | OFF | OFF |
| | B/R | OFF | OFF | | OFF | OFF | OFF |
| | R/B | 3-5kΩ | 3-5kΩ | 3-5kΩ | | ON | OFF |
| | B/W | 3-5kΩ | 3-5kΩ | 3-5kΩ | ON | | OFF |
| | W/Bl | OFF | OFF | OFF | OFF | OFF | |

**Nippon Denso unit** – set meter to K ohm scale

| Put negative (−) pin of tester to: | Put positive (+) pin of tester to: | | | | |
|---|---|---|---|---|---|
| | | R/W | B/W | B/R | B/Y | W/Bl |
| | R/W | | OFF | OFF | OFF | OFF |
| | B/W | 700-750Ω | | OFF | 400-450Ω | OFF |
| | B/R | OFF | OFF | | 400-450Ω | OFF |
| | B/Y | OFF | OFF | OFF | | OFF |
| | W/Bl | OFF | OFF | OFF | OFF | |

**Kokusan unit** – set meter to ohm x 100 scale

**Fig. 7.4 CDI unit test tables**

| | | | |
|---|---|---|---|
| B | Black | W | White |
| Bl | Blue | Y | Yellow |
| R | Red | | |

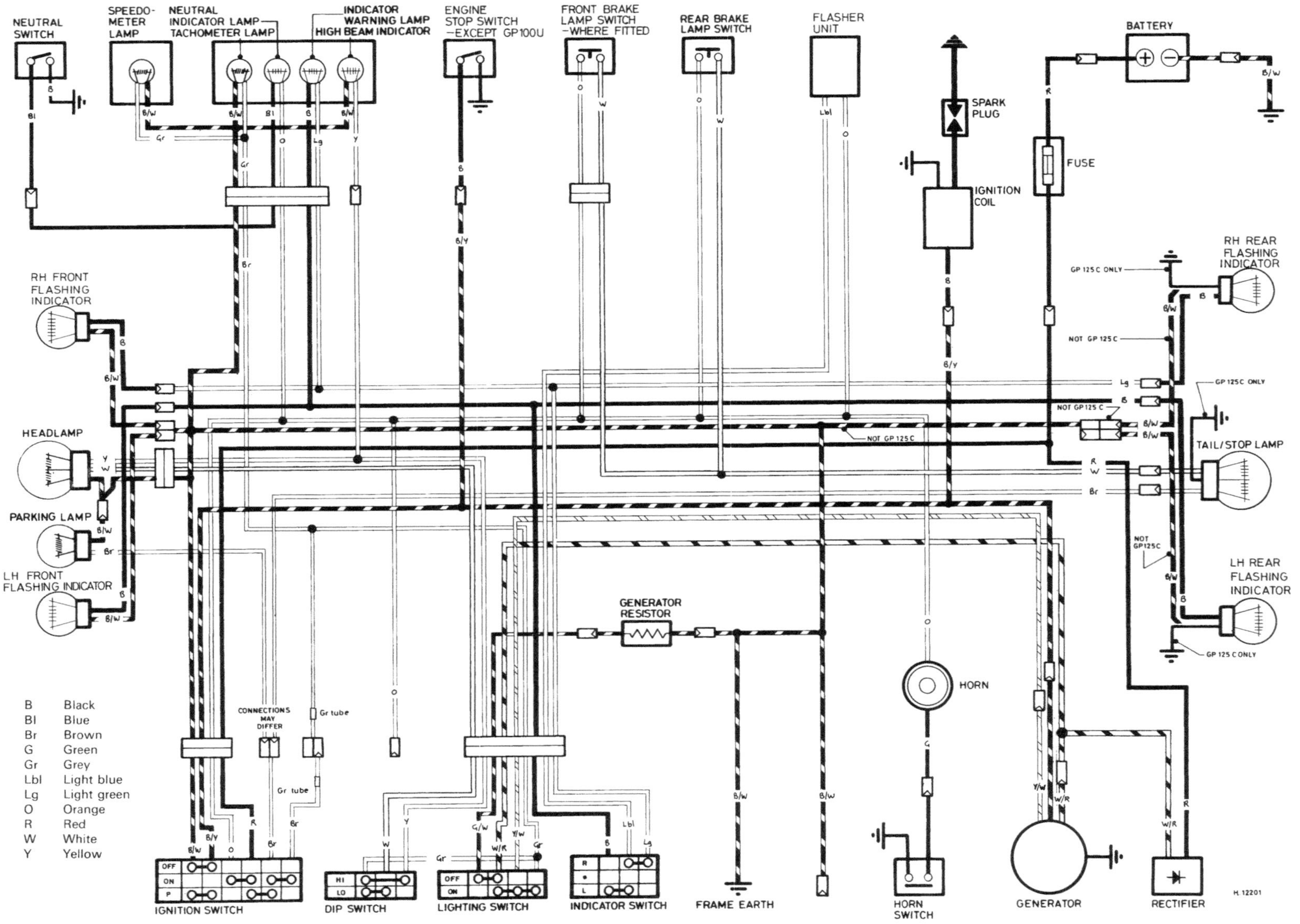

RH REAR FLASHING INDICATOR

TAIL/STOP LAMP

LH REAR FLASHING INDICATOR

GP 125C ONLY

NOT GP125 C

BATTERY

RECTIFIER

GENERATOR

FUSE

SPARK PLUG

IGNITION COIL

HORN

HORN SWITCH

FLASHER UNIT

FRAME EARTH

REAR BRAKE LAMP SWITCH

GENERATOR RESISTOR

INDICATOR SWITCH

FRONT BRAKE LAMP SWITCH – WHERE FITTED

LIGHTING SWITCH

ENGINE STOP SWITCH – EXCEPT GP100U

DIP SWITCH

INDICATOR WARNING LAMP
HIGH BEAM INDICATOR

NEUTRAL LAMP
INDICATOR LAMP
TACHOMETER LAMP

CONNECTIONS MAY DIFFER

IGNITION SWITCH

SPEEDO-METER LAMP

NEUTRAL SWITCH

RH FRONT FLASHING INDICATOR

HEADLAMP

PARKING LAMP

LH FRONT FLASHING INDICATOR

B    Black
Bl   Blue
Br   Brown
G    Green
Gr   Grey
Lbl  Light blue
Lg   Light green
O    Orange
R    Red
W    White
Y    Yellow

Wiring diagram – all GP100 and 125 models except the 1985 to 1986 GP125 N

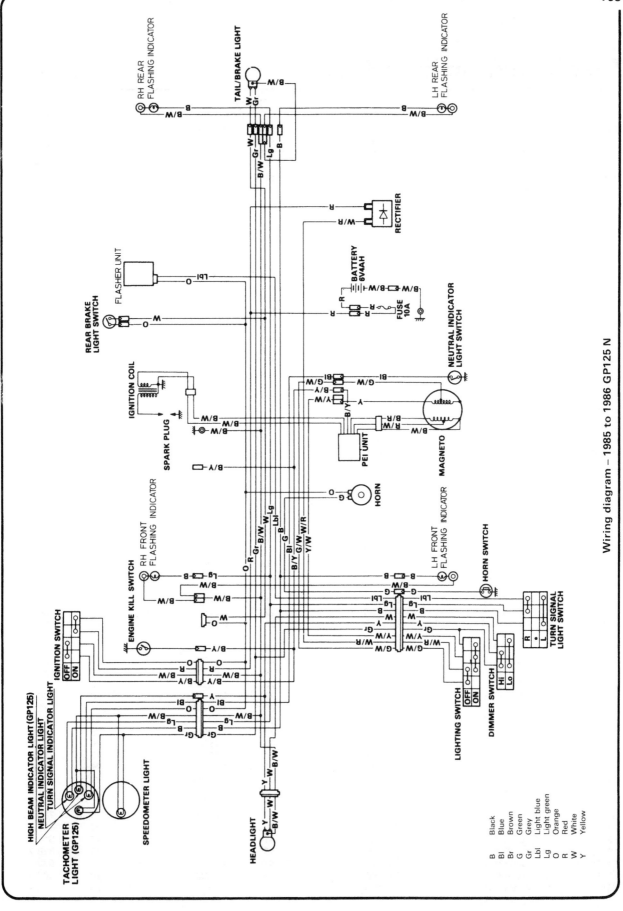

Wiring diagram – 1985 to 1986 GP125 N

# Metric conversion tables

| Inches | Decimals | Millimetres | Millimetres to Inches | | Inches to Millimetres | |
|--------|----------|-------------|-----|---------|--------|-----|
| | | | mm | Inches | Inches | mm |
| 1/64 | 0.015625 | 0.3969 | 0.01 | 0.00039 | 0.001 | 0.0254 |
| 1/32 | 0.03125 | 0.7937 | 0.02 | 0.00079 | 0.002 | 0.0508 |
| 3/64 | 0.046875 | 1.1906 | 0.03 | 0.00118 | 0.003 | 0.0762 |
| 1/16 | 0.0625 | 1.5875 | 0.04 | 0.00157 | 0.004 | 0.1016 |
| 5/64 | 0.078125 | 1.9844 | 0.05 | 0.00197 | 0.005 | 0.1270 |
| 3/32 | 0.09375 | 2.3812 | 0.06 | 0.00236 | 0.006 | 0.1524 |
| 7/64 | 0.109375 | 2.7781 | 0.07 | 0.00276 | 0.007 | 0.1778 |
| 1/8 | 0.125 | 3.1750 | 0.08 | 0.00315 | 0.008 | 0.2032 |
| 9/64 | 0.140625 | 3.5719 | 0.09 | 0.00354 | 0.009 | 0.2286 |
| 5/32 | 0.15625 | 3.9687 | 0.1 | 0.00394 | 0.01 | 0.254 |
| 11/64 | 0.171875 | 4.3656 | 0.2 | 0.00787 | 0.02 | 0.508 |
| 3/16 | 0.1875 | 4.7625 | 0.3 | 0.01181 | 0.03 | 0.762 |
| 13/64 | 0.203125 | 5.1594 | 0.4 | 0.01575 | 0.04 | 1.016 |
| 7/32 | 0.21875 | 5.5562 | 0.5 | 0.01969 | 0.05 | 1.270 |
| 15/64 | 0.234375 | 5.9531 | 0.6 | 0.02362 | 0.06 | 1.524 |
| 1/4 | 0.25 | 6.3500 | 0.7 | 0.02756 | 0.07 | 1.778 |
| 17/64 | 0.265625 | 6.7469 | 0.8 | 0.03150 | 0.08 | 2.032 |
| 9/32 | 0.28125 | 7.1437 | 0.9 | 0.03543 | 0.09 | 2.286 |
| 19/64 | 0.296875 | 7.5406 | 1 | 0.03937 | 0.1 | 2.54 |
| 5/16 | 0.3125 | 7.9375 | 2 | 0.07874 | 0.2 | 5.08 |
| 21/64 | 0.328125 | 8.3344 | 3 | 0.11811 | 0.3 | 7.62 |
| 11/32 | 0.34375 | 8.7312 | 4 | 0.15748 | 0.4 | 10.16 |
| 23/64 | 0.359375 | 9.1281 | 5 | 0.19685 | 0.5 | 12.70 |
| 3/8 | 0.375 | 9.5250 | 6 | 0.23622 | 0.6 | 15.24 |
| 25/64 | 0.390625 | 9.9219 | 7 | 0.27559 | 0.7 | 17.78 |
| 13/32 | 0.40625 | 10.3187 | 8 | 0.31496 | 0.8 | 20.32 |
| 27/64 | 0.421875 | 10.7156 | 9 | 0.35433 | 0.9 | 22.86 |
| 7/16 | 0.4375 | 11.1125 | 10 | 0.39370 | 1 | 25.4 |
| 29/64 | 0.453125 | 11.5094 | 11 | 0.43307 | 2 | 50.8 |
| 15/32 | 0.46875 | 11.9062 | 12 | 0.47244 | 3 | 76.2 |
| 31/64 | 0.484375 | 12.3031 | 13 | 0.51181 | 4 | 101.6 |
| 1/2 | 0.5 | 12.7000 | 14 | 0.55118 | 5 | 127.0 |
| 33/64 | 0.515625 | 13.0969 | 15 | 0.59055 | 6 | 152.4 |
| 17/32 | 0.53125 | 13.4937 | 16 | 0.62992 | 7 | 177.8 |
| 35/64 | 0.546875 | 13.8906 | 17 | 0.66929 | 8 | 203.2 |
| 9/16 | 0.5625 | 14.2875 | 18 | 0.70866 | 9 | 228.6 |
| 37/64 | 0.578125 | 14.6844 | 19 | 0.74803 | 10 | 254.0 |
| 19/32 | 0.59375 | 15.0812 | 20 | 0.78740 | 11 | 279.4 |
| 39/64 | 0.609375 | 15.4781 | 21 | 0.82677 | 12 | 304.8 |
| 5/8 | 0.625 | 15.8750 | 22 | 0.86614 | 13 | 330.2 |
| 41/64 | 0.640625 | 16.2719 | 23 | 0.09551 | 14 | 355.6 |
| 21/32 | 0.65625 | 16.6687 | 24 | 0.94488 | 15 | 381.0 |
| 43/64 | 0.671875 | 17.0656 | 25 | 0.98425 | 16 | 406.4 |
| 11/16 | 0.6875 | 17.4625 | 26 | 1.02362 | 17 | 431.8 |
| 45/64 | 0.703125 | 17.8594 | 27 | 1.06299 | 18 | 457.2 |
| 23/32 | 0.71875 | 18.2562 | 28 | 1.10236 | 19 | 482.6 |
| 47/64 | 0.734375 | 18.6531 | 29 | 1.14173 | 20 | 508.0 |
| 3/4 | 0.75 | 19.0500 | 30 | 1.18110 | 21 | 533.4 |
| 49/64 | 0.765625 | 19.4469 | 31 | 1.22047 | 22 | 558.8 |
| 25/32 | 0.78125 | 19.8437 | 32 | 1.25984 | 23 | 584.2 |
| 51/64 | 0.796875 | 20.2406 | 33 | 1.29921 | 24 | 609.6 |
| 13/16 | 0.8125 | 20.6375 | 34 | 1.33858 | 25 | 635.0 |
| 53/64 | 0.828125 | 21.0344 | 35 | 1.37795 | 26 | 660.4 |
| 27/32 | 0.84375 | 21.4312 | 36 | 1.41732 | 27 | 685.8 |
| 55/64 | 0.859375 | 21.8281 | 37 | 1.4567 | 28 | 711.2 |
| 7/8 | 0.875 | 22.2250 | 38 | 1.4961 | 29 | 736.6 |
| 57/64 | 0.890625 | 22.6219 | 39 | 1.5354 | 30 | 762.0 |
| 29/32 | 0.90625 | 23.0187 | 40 | 1.5748 | 31 | 787.4 |
| 59/64 | 0.921875 | 23.4156 | 41 | 1.6142 | 32 | 812.8 |
| 15/16 | 0.9375 | 23.8125 | 42 | 1.6535 | 33 | 838.2 |
| 61/64 | 0.953125 | 24.2094 | 43 | 1.6929 | 34 | 863.6 |
| 31/32 | 0.96875 | 24.6062 | 44 | 1.7323 | 35 | 889.0 |
| 63/64 | 0.984375 | 25.0031 | 45 | 1.7717 | 36 | 914.4 |

# Index